To
My uncle K. Srinivasa Adiga BA, MSc, DIISc
A tribute to his life-long pursuit of excellence in engineering

Intelligent Manufacturing

Manufacturing is awaiting a great challenge – the challenge of Artificial Intelligence (AI). We are witnessing the proliferation of applications of AI in industry, ranging from finance and marketing to design and manufacturing processes. The AI tools have been incorporated into computer-aided design software, shop-floor operations software, as well as in entering the logistics systems.

The success of AI in manufacturing can be measured by their growing number of applications, releases of new software products, companies developing and distributing these products, conferences and new publications.

The book series on Intelligent Manufacturing has been established in response to this great challenge.

The aim of the series is to publish high quality books on applications of artificial intelligence in manufacturing. The titles to be published in the series include topics such as Design for Manufacturing; Concurrent Engineering; Process Planning; Production Planning and Scheduling; Programming Languages and Environments; and Design, Operations and Management of Intelligent Systems.

Some of the titles are more theoretical in nature while others emphasize an industrial perspective. Books dealing with the most recent developments will be edited by leaders in the particular fields. In areas that are more established, books written by recognized authors are being planned.

The book authors, editors, and publishers certainly hope that the titles in the series will be appreciated by students entering the field of intelligent manufacturing, academics, design and manufacturing managers, system engineers, analysts, and programmers.

As the series editor, I would be most pleased to hear readers' reaction on the books published in the series.

Andrew Kusiak
Book Series Editor
Department of Industrial Engineering
The University of Iowa
Iowa City, IA 52242
USA

Object-oriented
Software for
Manufacturing Systems

Object-oriented Software for Manufacturing Systems

Edited by

S. Adiga

Assistant Professor
Industrial Engineering and Operations Research
University of California
Berkeley
USA

SPRINGER-SCIENCE+BUSINESS MEDIA, B.V.

First edition 1993
© 1993 Springer Science+Business Media Dordrecht
Originally published by Chapman & Hall in 1993

Typeset in by 10/12 pt Times by Pure Tech Corporation, India.

ISBN 978-0-412-39750-9   ISBN 978-94-011-4844-3 (eBook)
DOI 10.1007/978-94-011-4844-3

A catalogue record for this book is available from the British Library

Library of Congress Cataloging-in-Publication data available

Printed on permanent acid-free text paper, manufactured in accordance with the proposed ANSI/NISO Z 39.48-199X and ANSI Z 39.48-1984

Contents

Contributors

Sadashiv Adiga

Sadashiv Adiga has been an Assistant Professor (Industrial Engineering & Operations Research) at the College of Engineering, University of California at Berkeley since 1987. He earned a PhD in Industrial Engineering from Arizona State University. He also has degrees in Electrical Engineering and Business Management. Previously, he worked for Tata Engineering & Locomotive Co. (TELCO) in India for about ten years. His technical interests are related to systems integration in manufacturing organizations. He is a member of IIE, IEEE, SME, TIMS and AAAI

IEOR Department
4135 Etcheverry Hall
University of California at Berkeley
Berkeley, CA 94720
USA

Patrick Cogez

Patrick Cogez is currently Industrial Engineering Manager for the Crolles Plant of SGS – Thomson Microelectronics. Earlier he was deputy head of the Electricity Office in the French Ministry of Industry, where he was mainly in charge of investment and pricing studies. Patrick Cogez received his PhD degree in Industrial Engineering from the University of California at Berkeley in May 1991. He holds engineering degrees from Ecole Polytechnique and Ecole Narionale des Ponts et Chaussees in France, and received his MS degree in Operations Research from University of California at Berkeley in 1985.

ZI du pre Roux
BP 16
38190 CROLLES
FRANCE

Milind Gadre

Currently Milind is a Senior Software Development Engineer with Consilium Inc. in Mountain View, California. He is part of the team developing FlowStream – an Integrated Plant Floor Management System. Milind Gadre obtained a

Bachelor of Technology in Mechanical Engineering from the Indian Institute of Technology, Madras and a Master of Science in Mechanical Engineering from the University of Kentucky, Lexington. Milind also obtained the Master of Science in Industrial Engineering and Operations Research from the University of California at Berkeley. Technical interests include planning, scheduling and object-oriented software modelling of manufacturing processes.

Consilium Inc.
640 Clyde Court
Mountain View, CA 94043
USA

Jonathan Kolyer

He received his MS in Industrial Engineering and Operations Research from University of California at Berkeley and his BA in Mathematics from Northwestern University. Jonathan Kolyer's interests are in modelling and implementation of software systems, and writing on current trends in computing. He works as a software consultant and has recently completed projects for NASA Ames Research Center and Stanford University.

Computer Services Group
703 Market Street, Suite 1501
San Francisco, CA 94103
USA

Barry Lozier

Barry Lozier is a Software Project Manager for Consilium, Inc. He was a principal architect of the Flowstream product, and oversees the Database Interface and Product Integration teams. He attended Harvard University, and has been involved in software engineering for the past eight years. He is a member of ACM and IEEE.

Consilium Inc.
640 Clyde Court
Mountain View, CA 94043
USA

Adeel Najmi

Adeel Najmi is Project Leader for Advanced Manufacturing Engineering at NCR Microelectronics in Fort Collins, Colorado. He is responsible for developing applications for decision support in CIM, production planning, and scheduling. He is also a PhD candidate in Industrial Engineering at the University of California at Berkeley where he received a Master's degree in the same discipline. He holds a Bachelor's degree in Mechanical Engineering also from Berkeley. He is a member of IIE, AAAI, ASME and SME. His research interests include CIM, artificial intelligence, and Human Factors.

Advanced Manufacturing Engineering
NCR Microelectronics
2001, Danfield Court
Fort Collins, CO 80525
USA

(address till December, 1992)
IEOR Department
4135 Etcheverry Hall
U. C. Berkeley,
Berkeley, CA 94720
USA

David Wallace

David is currently on the new product development staff at Flexis Control, Inc. He began work at Flexis Control in 1989 where he has been responsible for building manufacturing and marketing applications using object-oriented methods. He started his career at General Motors supporting the Material and product Test Laboratories. David Wallace graduated from Michigan State University in 1985 with a BS in Computer Science.

David Wallace
Flexis Control, Inc.
3102 Diablo Ave.
Hayward, CA 94545
USA

David Wilczynski

David Wilczynski is currently President Of Pacific Software Solutions, Inc., which he founded in 1992. He co-founded Savoir (now Flexis Control, Inc.) in 1984, where he was responsible for the actor methodology. As part of the Department of Defense's Automated Airframe Assembly Program, he led a team applying advanced uses of the actor methodology to aerospace factory floor problems. Dr Wilczynski received his PhD from University of Southern California in 1975 specializing in Artificial intelligence. He taught there briefly and did research at its Information Sciences Institute. His work focused on large system design and knowledge representation.

Pacific Software Solutions, Inc.
1911 Ernest Ave. #A
Redondo Beach, CA 90278

Paul Worhach

Paul Worhach is currently a PhD candidate in the department of Industrial Engineering & Operations Research at the University of California at Berkeley.

He has worked as a management consultant and manufacturing systems consultant and most recently worked with SEMATECH on the design of object-oriented simulations of semiconductor manufacturing equipment. He received his BSE from Princeton University and his MS in Industrial Engineering & Operations Research from the University of California at Berkeley.

IEOR Department
4135 Etcheverry Hall
University of California at Berkeley,
Berkeley, CA 94720
USA

Preface

I must confess that I stumbled upon the object-oriented (OO) world view during my explorations into the world of artificial intelligence (AI) in search of a new solution to the problem of building computer-integrated manufacturing systems (CIM). In OO computing, I found the constructs to model the manufacturing enterprise in terms of information, a resource that is common to all activities in an organization. It offered a level of modularity, and the coupling/binding necessary for fostering integration without placing undue restrictions on what the individual applications can do.

The implications of OO computing are more extensive than just being a vehicle for manufacturing applications. Leaders in the field such as Brad Cox see it introducing a paradigm shift that will change our world gradually, but as radically as the Industrial Revolution changed manufacturing. However, it must be borne in mind that simply using an object-oriented language or environment does not, in itself, ensure success in one's applications. It requires a different way of thinking, design discipline, techniques, and tools to exploit what the technology has to offer. In other words, it calls for a paradigm shift (as defined by Kuhn in *The Structure of Scientific Revolution*, a classic text in the history of science).

Paradigm shifts need involvement from a variety of people, such as architects, engineers, accountants, and experts in individual domains, to perceive the benefits and get involved in this change. I have supplemented my thoughts and experience with those of other professionals who are pioneering OO applications in manufacturing.

My journey into the object world began in Arizona State University with encouragement from Richard Smith, who was then the Chairperson of the Department of Industrial and Management Systems Engineering. I delved further into it with active support and participation from my colleague Roger Glassey at the University of California at Berkeley. In 1987, Roger and I started work on the software object library, the Berkeley Library of Objects for Control and Simulation for Manufacturing (BLOCS/M). Much of my experience has come from BLOCS/M.

Many people have contributed to BLOCS since 1987. The list of people working with it is growing every year. However, I would like to make special mention of the 'original' team of 1987 which participated in the first set of experiments. The team consisted of Roger, myself, Christopher Lozinski, Woo-Tsong Lin,

Raja Petrakian and Jong-Soo Kim (the last four were graduate students in the Industrial Engineering and Operations Research Department in 1987). However, I acknowledge the contribution of all the participants to the BLOCS project (including the ones not mentioned).

Colleagues Ted Crossman and Robert Leachman have supported us in many ways, especially through invaluable criticism and feedback, on our journey into the object-oriented world. I would also like to thank Bill Jewell and David Dornfeld (past and present Directors of the Engineering Systems Research Center) who, along with their staff, helped us manage the resources to run the software projects.

I gratefully acknowledge the support of all sponsors of our research including the MICRO program of the University of California, Semiconductor Research Corporation, NCR Microelectronics (Fort Collins, CO), Harris Corporation, the Manufacturing Systems Research Group at the T.J. Watson Research Center of IBM Corporation, and the Stepstone Corporation. Special thanks go to Tim McCarthy and Dan Ellsworth of NCR Corporation, Fort Collins for trying out this technology early and providing us with valuable feedback.

I would like to thank all the people who contributed chapters to this book. Special thanks are also due to Mark Hammond of Chapman and Hall for his encouragement and patience during the preparation of this manuscript.

Finally, I could not have completed this book without the moral support from my best friend and wife, Rekha.

Sadashiv Adiga
Berkeley, California, USA
1992

Part One
Conceptual Background

1 Introduction

S. ADIGA

The use of computers in manufacturing brought many productivity gains in the 1980s, but modern manufacturing systems have become extremely complex to manage. The manufacturing scene is beset by constant changes in technology, product and production volume mix. Manufacturing industries have tried to automate production and its management at varying levels of control and financial investment with mixed success.

A popular view is that the full potential of productivity improvements made possible by modern electronic computing techniques is not being realized because of the lack of a systems approach capable of integrating the 'islands' of automation – which tend to proliferate with progress in manufacturing engineering – while allowing each such island to function to its fullest extent. This has led to the idea of designing systems with integration through computers as the main theme.

While many technologies have contributed to the success of individual applications related to CIM, no single technique or method has appeared to provide the basis for designing CIM systems in general terms. Based on what we have seen of the promise of object-oriented technology in the late 1980s, it has the potential to provide the conceptual and computing philosophy needed to achieve integration of the manufacturing enterprise. In addition to helping solve specific problems, it also has the potential to change methods of computing in other domains in the future.

1.1 Objects and object-oriented software

Object-oriented (OO) computing is a style of computing in which data and associated procedures are encapsulated to form an 'object' (Peterson, 1987). Thus, an object is a computational entity that exists at a higher level of abstraction than data structures or procedures. The term 'encapsulation' implies that the data can be accessed only through pre-defined procedures. The other main principle in this paradigm is that objects communicate with each other only by exchanging messages. The principles of OO computing can be applied to the design of both hardware and software. In this book, we limit our discussion only to software, therefore we use the term OO software (OOS) to describe the software designed according to the principles of OO computing.

1.2 Objective and target audience of the book

The main objective of this book is to interpret and apply OO concepts in the context of designing manufacturing systems applications. The main audience for this book consists of professionals (engineers and managers) dealing with manufacturing systems, students, and educators looking for new directions in building software systems to solve problems in this area. This book should be of special interest to engineering and computer professionals who have heard the term 'object-oriented', and want to learn more about it and its import-ance, particularly in designing software for manufacturing systems. This group includes:

- manufacturing and/or industrial engineers;
- software professionals;
- AI professionals; and
- teachers and students of manufacturing systems modeling and information systems.

Many aspects of object-oriented computing are discussed in this book. We pro-vide an introduction to the technology, design techniques, and discuss examples to illustrate the relevance of object-orientation in designing software to solve some important problems in manufacturing.

This book is not a cookbook of recipes. The focus is on providing an overview of different aspects of OO technology and how they can be applied in building software to represent and control manufacturing systems. Topics are covered in detail to support this overview; however, an implementor will need to supplement this book with others on programming. When you finish this book, we expect that you will be familiar with:

- the terms and concepts underlying object-orientation in software construction;
- the benefits of object-orientation in developing software for applications in manufacturing;
- some techniques used in the design and development process in the context of manufacturing systems;
- some important organizational issues affecting the use of OOS in manufacturing.

Additionally, the book contains a model of a new architecture on which to base new developments of OO software for manufacturing, and a new prototyping approach illustrated through an example.

1.3 Organization of the book

This book attempts to address a comprehensive range of issues, including issues related to languages, design techniques, conceptual to industrial examples, and management of technology. We have made a conscious effort to be consistent in the use of terminology with respect to object-oriented technology; a glossary of some prominent terms is provided at the end of the book.

The book is organized into five parts: conceptual background; design and implementation techniques; some examples of practical applications; management issues (from an organizational point of view); and, finally, conclusions and future directions. Within each section, the presentation varies from the general to the more technical in content. We adopt this approach as we expect most readers to prefer a 'general' to a 'specific' approach when discussing a new technology.

Part One provides the conceptual background needed to understand the material covered later. Chapter 2 is an introduction to the basic ideas underlying object-orientation in software. Chapter 3 contains a presentation that relates these ideas to specific issues in building software for manufacturing systems.

Part Two contains a discussion of the techniques/tools and ideas needed to systematically develop computer-integrated manufacturing (CIM) applications. Chapter 4 has a proposal for a new architecture for application in a CIM environment. A prototype of this architecture is being built at Berkeley. An approach found to be useful in prototyping this work is described in Chapter 5. It contains a conceptually simple technique that builds on the currently popular techniques in data modeling. Chapter 6 focuses on OO databases. Here, the readers are assumed to be familiar with the basic database concepts, particularly the relational model. Readers interested in reviewing relational database concepts may choose from a variety of books including the one written by Date (1990). Chapter 7 is on languages with a special focus on C-based OO languages. The reason for picking derivatives of C is that, in manufacturing, the control of processes in real-time proves to be the central issue, and C-based languages are the most promising in this area.

A link between Parts Two and Three is provided by the organization of the material; Part Two is organized in a top-down fashion; i.e.:

1. OO systems architecture;
2. design of object libraries;
3. OO data management;
4. OO languages and object-oriented programming.

Part Three also reflects this organization:

1. requirements leading to high level systems architecture or specification;
2. object library design/composition, design issues;
3. programming issues;
4. implementation, system operation, etc.;
5. maintenance etc.

Two different industrial applications are discussed in Part Three. These chapters reflect viewpoints of different classes of users of this technology. Together the chapters in Part Three present examples of use of OO technology by someone involved in the development of a commercial product, and by people using a commercial product to design two applications for a client. Chapter 8 is about the design and implementation of a new product for factory floor management. Barry Lozier of Consilium describes the details of development and implementation of a new commercial product. Many people have said that they believe that OO is real only when they see real-time control applications in manufacturing. David Wilczynski and David Wallace of Savoi discuss two different types of real-time control applications in chapter 9.

No perspective on any technology is complete without a discussion of management/organizational issues raised by the technology. In Part Four, Chapter 10, Adeel Najmi of NCR expresses his views as an engineer responsible for the management of this technology in his organization.

Concluding remarks are made in Part Five, Chapter 11. This chapter includes a discussion on object-oriented distributed systems, an important topic for future applications in industry and research.

At the end of Parts One, Two and Three, a summary is provided which highlights the main issues (from a systems perspective). These summaries are also intended to lay the groundwork for the material in the following section. A brief list of reading material for further study is also provided for interested readers.

A glossary is provided at the end of the book to clarify some of the terminology used, and a list of resources is included in the appendix. This does not claim to be an exhaustive list; but it is representative of the resources available for people who wish to keep in touch with new developments in this area.

1.4 Conclusion

Dreyfus and Dreyfus (1986) suggest that wishful thinking has always complicated our relations with technology. This has, perhaps, led to unrealistic expectations from some technologies as we have seen in the case of artificial intelligence.

Object-orientation in computing means different things to different people. Some people see the benefits from OO computing as merely an addition of an abstract data type to programming facilities, whereas others like Jim Waldo (1989) see in it the emergence of a new generation in computing. In our view, the benefits one may gain from any technology are directly related to his/her perception of it. To make a point using an extreme example, one can drive a helicopter along a highway and complain that it is too expensive a replacement for automobiles! On the other hand, one should avoid trying to travel to the moon (in a helicopter)!

The material in this book is influenced by our past work and future vision for OO thinking. While we believe that one may gain much by following the concepts/techniques discussed in this book, one should not overlook the fact that these ideas complement the fundamental skills of discipline and knowledge needed in good software design and implementation.

References

Date, C. J. (1990) *Introduction to database systems* 5th edition. Addison-Wesley, Reading, MA.

Drefus, H. L. and Dreyfus, S. E. (1986). *Mind Over machine*. The Free Press, New York.

Peterson, G. E. (1987) *Tutorial: Object-oriented computing*. (Preface). IEEE Computer Society Press, Washington D.C.

Waldo, J. (1989) A new generation. *Unix Review*, **6** (8), 33–40.

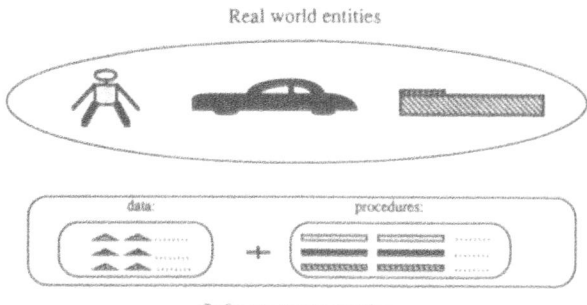

Figure 2.1 *Conventional representation.*

procedure (or method) defines the behavior expected of the object. One such behavior is to change (or return) the data stored in its instance variables. Objects interact by sending one another messages. That is, an object sends a message to another when it wants certain service from the object receiving the message. Typically, receipt of a message activates a corresponding method in the receiving object.

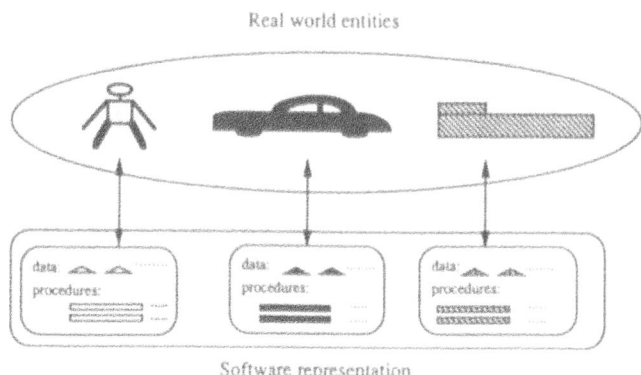

Figure 2.2 *Object-oriented representation.*

Sending a message to an object is similar to asking it to perform an operation on itself and return the result to some place or to the requesting object. From an execution point of view, sending a message is similar to calling a function in languages such as Fortran or C, with one important difference. A function in Fortran or C invokes a predetermined procedure whereas, in OOS, the decision of which procedure (or method) is invoked is left to the object receiving the message. Further, when a message tells a workstation object to update its state, it contains no information about how to execute the message.

Representation as objects is not necessarily limited to physical entities. Activities, events and relationships can all be represented as objects if they have information that is to be stored or manipulated to benefit the users.

2 Object-oriented software systems: Concepts

S. ADIGA

2.1 Introduction

The purpose of this chapter is to introduce the readers to some basic concepts of object-orientation in software systems. Additionally, we also present the benefits and limitations of this approach in developing software systems.

Software systems that implement object-oriented techniques are popularly called object-oriented programming systems (OOPS) or object-oriented software (OOS). Some representational aspects of today's OOPS can be found in an early simulation system, SIMULA (Dahl and Nygaard, 1966) and artificial intelligence systems using frames (Minsky, 1975). Object-oriented computing as understood and practised today owes its popularity to the ideas made popular among the software engineering community by the software Smalltalk (Goldberg and Robson, 1983). The increasing popularity of the OO concepts in recent years has led to the emergence of a new paradigm.

The OO paradigm is an outcome of the evolution of modularity concepts developed to improve various aspects of the software development life cycle. Earlier efforts in modularization led to the development of ideas such as the use of a library of functions, sub-routines, data files and abstract data types, among others. Most of these developments retained the traditional fundamental notion that

$$(\text{computer}) \text{ programs} = \text{data} + \text{algorithms (procedures)}$$

A basic problem with this approach is that most real-world entities are encapsulations of data and procedures that characterize their behavior. Therefore, a programmer or analyst must go through a transformation and often a restructuring process to construct programs to model the entities.

The OO paradigm represents a different way of looking at the program modules. It defines program modules as a package of data and procedures named an 'object': i.e. an abstraction of private data and operations that are naturally associated together. Because of this abstraction facility, we can represent real-life factory objects such as a machine or a part quite close to reality (Figure 2.1).

Data is stored in locations, i.e. instance variables, which cannot be directly accessed by other objects. Procedures are commonly known as 'methods'. Each

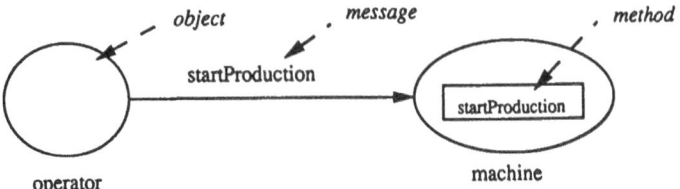

Figure 2.3 *Objects, a message and a method.*

2.2 Key concepts

Benefits to be gained from object-based computing are based on the concepts of classes, encapsulation, and polymorphism. These are supported and enhanced by important implementation or design techniques of inheritance, information hiding, and dynamic binding respectively. Each of these concepts and techniques will be discussed next.

2.2.1 Classes

A class defines the properties or behavior of similar objects by specifying the type of data and methods they share. Once a class is defined, any number of objects (called instance objects) may be created with the same structure, but with different particulars. This is done by repeatedly sending a message to the class requesting creation of an instance. Thus a class (or class object) may be viewed as a template to produce instance objects. A class may also serve to provide common instance variables and methods for a set of 'subclasses'. The concept of subclasses is discussed in the subsection below, 'Inheritance'.

Consider the example of a typical production system. In simplest terms, the purpose of a production system is to transform a set of raw materials into a set of finished products. Production systems typically use many different kinds of machines in this conversion process. The sequence of steps needing to be performed are generally found on a 'routing sheet'. Consider a part that requires processing on a lathe, milling machine, and a grinding machine before it is a finished product. One can identify a common property in these machinery as they require a capability to perform an operation (or step) that appears on the production routing sheet. Common behavior expected includes responding to signals to start or stop production.

We may describe 'Equipment' as a *class* of objects that represents the machinery involved in converting the raw material for this part to a finished product. A specific machine, namely, a grinder (of make, Norton) is called an instance (or member) of this class.

Inheritance

Inheritance is a powerful concept based on the ability to define subclasses derived from parent classes. One creates a software 'class' object by describing the most

common features/behavior required of a class or group of entities of the system to be modeled. When subgroups or subclasses are decided upon, they are defined by describing the specialization or any additional functionality needed in the respective subclasses.

Considering the earlier example, we looked at one part and its processing requirements. When all the parts to be made in that fabrication shop and their processing requirements are considered, we may decide that there are different types of machines that require special consideration. For example, parts are loaded into furnaces in a batch, whereas the gear-grinder used may handle individual parts one at a time. Furthermore, set-up and auxiliary equipment requirements are also considerably different. Therefore, we could define further specialization of the class Equipment in Figure 2.4.

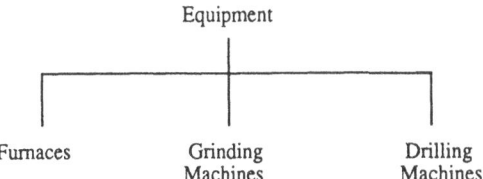

Figure 2.4 *A class and subclasses.*

A subclass inherits all the instance variables and methods from its superclass. This results in less coding if the inherited methods and instance variables are valid for the subclass. Additionally, a subclass can define its own instance variables and methods. By reimplementing an inherited method in a subclass, it may override the inherited method, if needed.

Inheritance is also a useful mechanism for broadcasting change. For example, if the factory (containing the classes Equipment, Drill, etc., as described earlier) decides to automate the collection of equipment status data, all that is needed is to describe this fact in the parent class, Equipment, and the rest of the subclasses inherit this change.

This concept of inheritance is the key to programmer productivity through reuse of software (objects). It eliminates the need for rewriting code and maintaining code in different places. It is a convenient mechanism to broadcast change, if needed, to all the subclasses. Some OO languages allow multiple inheritance, i.e. an object can inherit code from more than one superclass.

2.2.2 Encapsulation

The feature of confining all relevant data in a module and restricting its access to a set of pre-defined procedures is commonly known as encapsulation. It provides an abstraction to attain a natural one-to-one correspondence with real world objects. This property allows programmers to easily build and name software objects after their counterparts in the real world (Figure 2.2).

Behavior of objects is defined by their responses to messages. Objects use 'methods' to respond to messages. Since the only way an object reacts to the external world is through its response to the messages received, the methods provided inside an object form, in effect, the interface of an object with its external world.

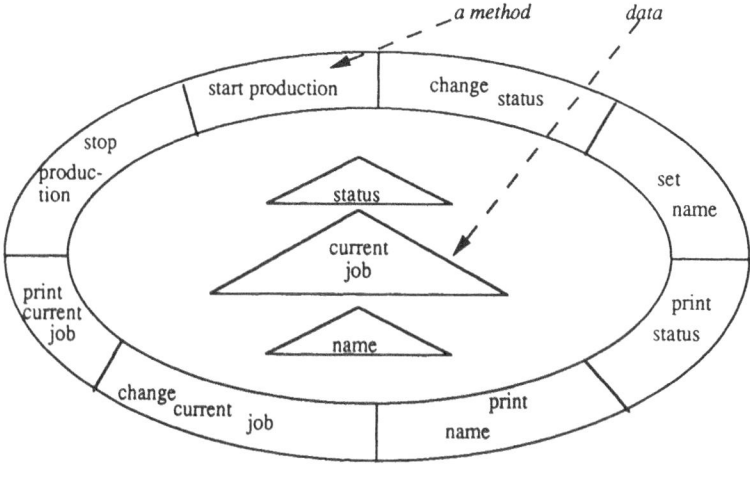

Software object - machine

Figure 2.5 *Methods acts as interface.*

Information hiding

Information hiding is an important design principle advocated by Parnas (1985) for handling the complexity of organizing software. He describes the principle of information hiding (in the context of system decomposition): 'System details that are likely to change independently should be the secrets of separate modules; only assumptions that should appear in the interfaces between modules are those that are considered likely to change.'

Information hiding implies that the implementation details of a module are hidden from the user. The user can only access an object through its declared interface. Furthermore, a potential user need not know how the object prepares its response. This allows a designer to change an object's internal details of implementation without affecting other parts of the system it is interacting with, as long as the interface remains the same.

Proper understanding of this property of 'information hiding' promotes good OO design. In simpler terms, you do not tell an object 'how to implement a service'; you only tell an object 'what to do'. Such an approach minimizes coupling between the sender and receiver of a message.

This restriction on accessing data hidden in objects, or allowing access to an object only through a protected interface, also reduces the possibility of accidental corruption of data and therefore makes the programs more reliable.

2.2.3 Polymorphism

Polymorphism means an ability to take more than one form. In software objects, it is the ability of different objects to respond differently to the same message in ways unique to their respective behavior. Figure 2.6 shown below illustrates this property.

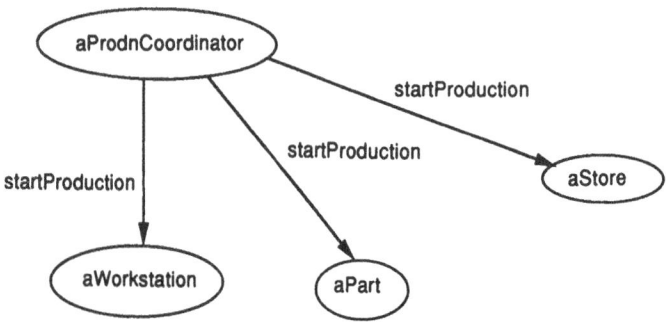

Figure 2.6   *Example of polymorphism.*

Figure 2.6 shows a sequence of messages triggered by an event indicating the start of processing of a part at a workstation. The objects involved are:

Object name	*Purpose*
aProdnCoordinator	– initiates production activities
aWorkstation	– represents a set of identical machines
aPart	– represents the part to be processed
aStore	– stores parts to be released to the line.

Three different objects – aWorkstation, aPart, aStore – receive the message 'startProduction' and react to it in their own ways:

aWorkstation	– decrements its count of idle machines and increments its count of busy machines
aPart	– changes its state from 'waiting' to 'busy'
aStore	– prepares to release a new part to the line.

Additionally, the above example also illustrates how polymorphism is exploited to convey the fact that the same event is experienced by several objects.

Proper combination of the properties of polymorphism and information hiding enables us to design objects that are interchangeable or 'plug-compatible'. One can easily plug new objects into the system if they respond to the same messages as the present ones. In Figure 2.3, consider that the machine is to be replaced by a new machine with, say, more capacity. Since the new machine has an implementation of the method, startProduction, the old object can be substituted with the new one. Use of polymorphism allowed different objects to respond to the same message description in individual ways; since no assumption was made on the implementation of the method, the new object could reimplement it without affecting the interface.

Dynamic binding

Binding is the term commonly used to refer to the process of functionally integrating a procedure and the data on which the procedure is required to operate. In most conventional languages, this decision is made when the code is being compiled. This process is known as early binding. In a language such as Smalltalk (which is a popular OOP language), this decision is made when the program is running. It is known as late or dynamic binding.

Dynamic binding relates to the process of associating a reference (message name) with a routine that performs the required function at the time it is being referenced. That is, the relationship between a message and the correct procedure is established when the message is actually issued. Dynamic binding offers a designer flexibility in many ways: since references to objects are symbolic (i.e. through their names), a method can be recompiled without having to recompile all of its callers (Duff, 1986).

Use of dynamic binding in combination with polymorphism allows different responses to the same message depending on the particular object instance receiving the message. Consider the situation where a finished part is to be picked up by one of many different material handlers. The object that sends the message 'movePart' need not know in advance, which material handler (say, one of the various types of Automated Guided Vehicles) will pick up the part that is ready to be moved. Since many different object classes may implement the message in their own unique ways (thanks to polymorphism), and the binding between the message call and method invoked is done during runtime, the system design can be quite flexible.

Use of dynamic binding in one's system design allows a system to adapt dynamically to changes in (system) behavior without recompilation. This feature is extremely useful in some process control situations where a system shut-down may be very expensive.

However, one must be cautious in using dynamic binding in languages such as Objective-C. Some type mismatches are not discovered until runtime. For example, in Figure 2.6, if the object, aPart, does not have a method to respond to the method, startProduction, this will be discovered in a runtime error. Languages like Objective-C provide warning messages during compilation.

2.3 Object communication

In this section, we discuss some important aspects of communication among objects.

2.3.1 Inter-object communication

The simplest model of communication between objects involves two objects where the sender of a message needs to know the identity of the receiving object, but not vice-versa. However, the information flow may be bidirectional, i.e. the receiving object may return information of interest to the sender. This return may be in the form of a value or result (similar to function return values) sent automatically by the receiving object. Alternatively, the receiving object may send a reply message (similar to handshaking).

There are two ways objects communicate with each other:

1. An object sending a message to another object (receiver) waits until it receives a result before proceeding with other messaging activity. This mode of communication is known as synchronous message passing.
2. The sending object continues execution immediately after sending a message. This is known as asynchronous message passing.

Variations of synchronous message passing include:

1. The sender just waits for an acknowledgement (not the expected result).
2. The message is aborted if the receiver is not ready.
3. The message is aborted if the receiver is not ready after some specified period.

The object model of communication through messaging does not make any distinction between messaging among local or remotely located objects. This allows one to design distributed systems as well as non-distributed systems using the same abstractions.

2.3.2 Concurrency and synchronization

Most discussions on object systems assume that messaging is synchronous. Concurrency is an important aspect of many real-time control applications in manufacturing. Concurrent systems consist of independent activities (or processes) that must communicate and synchronize to achieve some common goal.

Two different methods are used in handling concurrency and synchronization issues in OO approaches. In the first one, an object management system controls and synchronizes the access to objects. Thus individual objects may be regarded as 'passive' objects. This view is made popular by database implementations. In the other approach, the objects are 'active'; they replace the notion of a 'process' (Tsichritizis and Nierstrasz, 1989). With active objects, there is no need for an explicit synchronization mechanism as the objects themselves decide when they are ready to receive a message. Active objects are discussed in subsection 2.3.3.

In many OO systems, where there is a need for concurrent computational activity between objects, messages are sent to a synchronization entity. The latter then processes them serially. Objects are also free to implement their own synchronization mechanism internally for 'within' object synchronization, if needed.

In its simplest form, concurrency control may be achieved by simulating multiple processes (or tasks) in a single address space. A scheduler (provided by the operating system) manages these multiple processes by providing each process with a share of CPU time. All the tasks may communicate between each other and share computing resources voluntarily. They are also not protected from each other.

This simple system can be enhanced with algorithms for prioritizing, synchronizing and controlling processes. Most programming languages supporting concurrency provide special facilities, such as semaphores, monitors, coroutines, or the rendezvous, which enable the user to handle the necessary synchronization conditions (Atkinson *et al.*, 1991).

2.3.3 Active objects

An active object is one that has an independent thread of control, i.e. it has control over the execution of computation required. It can monitor events that occur during an event and take action autonomously. This allows an asynchronous behavior at the program level. These notions are formalized in the Actor models or languages.

It may be noted that the term Actor was introduced by Carl Hewitt (1977) to describe the concepts of reasoning agents. A detailed discussion on the underlying model of computation is found in Agha (1986); and a language that extends the basic concepts of Actors is discussed by Yonezawa (1990). Wilczynski (1988) discusses an approach for applying the Actor formalism to manufacturing applications. A practical example of this idea is discussed by Wilczynski and Wallace in Chapter 11.

2.4 OO analysis and design v. structured analysis and design

Many people were involved in the development of techniques that have come to be identified as structured analysis (SA) and structured design (SD): Larry Constantine, Tom DeMarco, Edward Yourdon, to name a few. SA and, later, SD popularized the idea of using data flow diagrams for requirements analysis and software design. Readers are referred to Yourdon (1990) for a comprehensive treatment of SA and SD. Though object-orientation in software is fundamentally different in some ways, it is also similar in its basic concepts to the fundamentals of SA and SD: these techniques also emphasize the principles of structure, orderly partitioning, systematic development, and independent modules.

Solving problems in the OO paradigm involves a different way of thinking about the world (as compared to conventional ways). OO software systems view the world as a collection of discrete objects which act and react in a common environment. The architecture of an OO software system is built around the object classes that represent the behavior of the entities in the system. Therefore the most important work in the design of OOS is the decomposition of the target system into object classes or set of objects that are to be manipulated. This is quite different from the conventional way of decomposing a system based on the functions to be performed by the software system.

A result of object-wise decomposition is that when the characteristics of entities such as furnaces are to be modified, they are done in only one place, i.e. where the object class Furance is defined. In the other approach, one has to examine all the functional modules, as the behavior of a furnace is spread over all the modules that use furnaces.

Once classes are defined in a problem domain, they are used repeatedly in building new software in the domain. For example, once we define the classes Equipment, Etcher, Furnace etc., any new application in the factory will be developed using these classes. The result is a different approach to software development: it is application development by the assembly and refinement of existing proven classes, rather than by writing code from scratch.

Another important difference is that in SA/SD, the analysis and design phases are distinctly separated. This separation is weak at best when building OO software. There is much overlap between the two phases.

2.5 Software life cycle benefits

A major goal of many users who adopt OOS is the use of its potential for a reduction in total software life-cycle cost (Buzzard and Mudge, 1985). Results of favorable programming experiences are documented by Tesler (1986) and Cox (1986), among others. In this section, we briefly discuss how this technology affects the typical life cycle of software.

The life of a software starts with systems analysis, resulting in a specification of requirements. This analysis serves as the basis for design. Design usually proceeds in two phases: general or high-level design, and detailed design. The design (of software) is then implemented and tested. After passing the tests, the software attains the 'product' status. During its use, it needs to be maintained with bug-fixes and extensions. To create an improved version of the software, the phases of specification, design, implementation and test have to be repeated. This traditional description is called the 'waterfall' model. The waterfall model views the software development process as a set of discrete, linear series of activities.

The life cycle of an OOS application may not be too different from that of other software developed using rapid prototyping. But the life cycle of a set of software object classes is different from the software constructs used in conventional programs, as these may be reused in other applications.

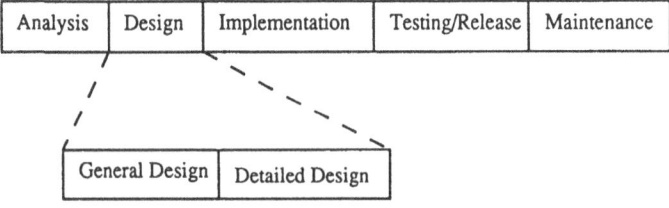

Figure 2.7 *The generic systems life cycle.*

The impact of object-orientation during some important stages in the software life cycle, as shown in Figure 2.7, is discussed next, while prototyping concerns are discussed in section 2.6.

2.5.1 Systems analysis and design

Tom DeMarco (1978, p. 4) defines 'analysis' as the study of a problem prior to taking some action; in the domain of computer systems development, he says, it refers to the study of some business area or application, usually leading to the specification of a new system.

While the focus of the systems analysis stage is on the user, the design stage focuses on finding good representations to meet user requirements. Usual approaches to design involve progressive decomposition toward more detail (Henderson-Sellers and Edwards, 1990). System design is often done in two phases: general and detailed design, or sometimes, logical and physical design.

Though the boundary between the analysis and design phases is not very rigid, the emphasis lies on different issues. In our view, this separation of analysis and design becomes even more fuzzy in OOS systems.

The property of encapsulation makes it is possible to have software objects that directly correspond to the physical entities in a manufacturing system such as machines, operators, lots etc. Therefore, the entities that the users and software engineers discuss, when defining the requirements as a part of the project, are the same entities designers build objects from and programmers work with during implementation of the requirements. This helps to make a smooth transition from requirements to design to implementation.

Compared to conventional approaches, a larger portion of time is spent on designing, because the software is being designed for easy reuse and future extensions. Another side-benefit of this approach is that it encourages more time to be spent in thinking about the conceptual design of the system. We have to look for similarities among objects if we are to take advantage of the inheritance mechanism. The use of OO programming tools does not guarantee good software!

The use of objects and messages allows an easily understandable system design to be developed. It is our experience (Glasscy and Adiga, 1989) that one can design software objects that correspond closely to the real-life physical objects in the area of manufacturing. This results in a more accurate representation of the real world in OOS.

2.5.2 Implementation

Since program development follows the abstraction process, it allows software objects to be developed in parallel after the interfaces are defined. This is possible because the implementation details of one object are independent of other objects. This advantage is further supported by the fact that objects can be separately compiled in OOP languages; thus, a large problem can be broken into manageable chunks and given to different programmers for implementation.

As mentioned earlier, object-oriented design makes implementation easier by improving the task of communication between designers and implementors. In principle, object-oriented thinking can be implemented in any programming language. But it is done most conveniently in a language or environment already designed or extended to provide for the concepts needed to support the basic features discussed earlier. The selection of a language should be on the basis of the appropriateness and the expectation of the use of the OOP facilities it provides.

An overview of a few of the major languages with special emphasis on a comparison between extensions to the C language appears in Chapter 7.

2.5.3 Testing

A major goal of testing is to demonstrate that the code or program has produced the functionality specified. It involves testing the individual units as well as the system as a whole. Since an OO application contains clearly defined and separately identifiable modules, these can be tested one at a time. Separate testing of individual objects before they are put into the system helps to localize errors.

2.5.4 Maintenance

Encapsulation restricts any undesired side-effect from changing the contents of any object's data. Since all the data and procedures related to an object are located in one place (rather than being spread all over the program), changes to be made are confined to one location. Apart from preventing any accidental corruption of data, this feature helps in both the maintainability and the extendibility of software. Moreover, according to Tracz (1987), software components designed for reuse generally have a very low defect rate.

Use of explicit communication through messages and polymorphism allows the use of entirely new classes of objects in an existing application, as long as they follow the same message protocol as the application.

According to Byard (1990), development and maintenance problems with current software techniques have created a large software backlog. He adds that about half this backlog is due to changes in user requirements. Another 25% is due to changes in code/data representation and emergency changes. Under an OO design, only the internals of an object can be changed without affecting the rest of the software. This feature is quite helpful in implementing changes.

2.6 Prototyping and software evolution

Prototyping is the process of developing a scaled-down version of a system to be used in building a full-scale system (Tanik and Yeh, 1989). In the engineering professions, prototyping has long been used to support analysis activity for building an operational requirements model. This is then used by the analysts, designers, testers etc., as the basis for building the product. Prototyping is gaining in popularity among software professionals as an approach to the understanding of problems involved in a complex or large project.

The flexibility offered by an OO approach presents some special advantages for prototyping. Since one object can be treated like any other as long as the two have the same message protocol (the idea of polymorphism), we can build large complex systems from smaller, interchangeable ones. But unlike the conventional approaches, the initial prototype need not be thrown away: it can be 'grown' into the full production system. Each prototype may be treated as an internal release followed by a customer release for review. This incremental development approach is also good for the morale of the developers who see a working system early in the product's life. For customers (of software), it provides a working version to test important ideas and to give feedback to the designers.

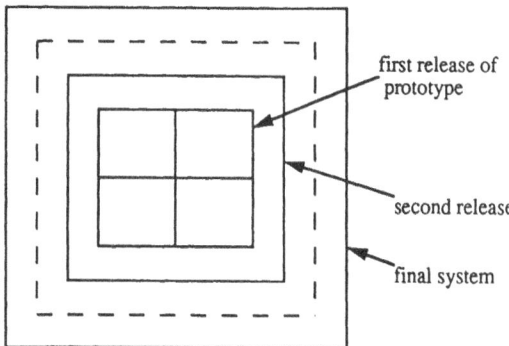

Figure 2.8 *Prototype grown to be the final system.*

The blocks that make up the prototype only provide minimum functionality of all the essential blocks that make up the final system. Typical examples of blocks in a production management system are: production scheduling, part tracking, and process control modules. All the modules need to exist in a minimum functional mode to provide a sense of an integrated system. The system is integrated right from the implementation of the first prototype. As a result, the problem of integrating many independently-built modules at the end of the project does not occur. The scope of the modules in a prototype is set by the user demands selected to be tested in the prototype.

The overall process of developing software using this approach is represented in Figure 2.8. The 'build and grow prototype' process encompasses aspects of design, testing and release activities described as traditional software life cycle activities. A method of developing prototypes using an OO approach is described in Chapter 5.

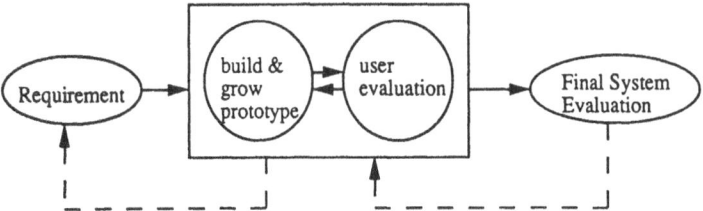

Figure 2.9   *OO system development through prototyping.*

2.7 Software reuse

The basic OO concepts and implementation techniques support the development of software objects that can be reused in more than one application. Such reuse is one important reason for an improvement in the productivity of programmers using OOP environments. Reuse of objects occurs when the same objects are used in different applications, or when new objects are added or 'plugged-in' to an existing application.

Software reuse does not occur by accident (Winblad *et al.*, 1990), even when an OO programming language is used. Proper design significantly enhances the reusability of software objects. Meyer (1988) discusses the role of design in promoting software reuse. Polymorphism and inheritance are among the main features of object-orientation that promote software reuse.

2.7.1 Reuse through polymorphism

Protocol for an object is defined by the set of messages to which it can respond. Objects which have an identical protocol are interchangeable. If different object classes contain the same protocol, they can be considered as 'plug-compatible'. Inheritance dominates such situations; it can also be achieved by using polymorphism. For example, one can access an Objective-C array in which each element is of different class. Since all the elements can respond to the messages being sent to them, any object can interact with the elements without worrying about their respective classes. Such a feature can be used to display different types of objects (windows, icons, etc.) by sending a common message – 'display' – to the elements in the array. The elements can be replaced by any other set of elements, if the new set implements the message 'display'.

2.7.2 Reuse through inheritance

There are two ways to promote reusability through the class hierarchy concept (Pinson and Wiener, 1991). The first one is to design a hierarchy of base classes to which subclasses may be logically added. Classes at a higher level tend to be more general than the lower levels. An example is that of the Equipment and Furnace classes presented earlier. Special types of furnaces may be added as subclasses to the Furnace class.

The second approach is to identify types of objects that are common to a large group of problems. The hierarchy of classes, Object–Collection–Sequenceable-Collection–ArrayedCollection–String from the language Smalltalk is a good example of this approach.

2.8 Potential limitations

Although the technology is promising, it is not without limitations. It is our observation that many people, particularly experienced programmers, have problems adjusting to thinking in terms of objects and messages. While the reasons for this may be debated, it is an observation shared by many other people who have managed OO projects.

It is also a fact that the promised productivity improvements through reusability start only after a library of software objects has been built or bought. Building reusable objects is not an easy task; furthermore, one must learn the library well before doing any serious programming. This makes the need for good documentation very important, and experience has taught us that good documentation is not something we can automatically expect with good software.

Languages such as Smalltalk are known to demand more machine resources than the conventional procedural languages. One reason for this is that most OOP languages provide more services than the conventional languages. It is also true that most of the concerns about execution speed are based on the performance of early research systems that were not designed for speed (Taylor, 1990).

Currently, there is also a shortage of personnel, both programming and managerial, who are conversant with this technology. However, this limitation may not remain for long, as object-orientation in computing systems has been receiving much attention in recent times.

The OO software industry suffers from the other drawbacks common to a young industry. That is, there is an absence of standards to enforce consistency among applications. The emergence and active role in promoting application standards by the OMG (Object Management Group), an industry consortium, is encouraging.

2.9 Conclusion

Object-oriented programming is one of the most talked-about topics in software engineering today. This technology offers hope to reduce the total life cycle cost of software development. Among its major advantages are improved programmer productivity, software reliability, and provision of data structures to model reality more accurately. Manufacturing engineers involved in complex software development have a lot to gain by investigating this approach.

References

Agha, G. (1986) *Actors: A model of concurrent computation in distributed systems.* MIT Press, Cambridge, MA.

Atkinson, C., Goldsack, S., D; Maio, A. and Bayan, R. (1991) Object-oriented concurrency and distribution in DRAGOON. *Journal of Object-oriented Programming*, **4**(1), 11–18.

Buzzard, G. D. and Mudge, T. N. (1985) Object-based computing and the Ada programming language. *IEEE Computer*, March, 11–19.

Byard, C. (1990) Object-oriented technology: A must for complex systems. *Computer Technology Review*, Fall, 15–19.

Cox, B. J. (1986) *Object oriented programming.* Addison-Wesley, Reading, MA.

Dahl, O. J. and Nygaard, K. (1966) SIMULA: An Algol-based simulation language. *Communications of the ACM*, **9**, 671–8.

DeMarco, T. (1978) *Structured analysis and system specification.* Prentice-Hall, Englewood, NJ.

Duff, C. B. (1986) Designing an efficient language. *BYTE*, August.

Glassey, C. R. and Adiga S. (1989) Design of a software object library for simulation of semiconductor manufacturing systems. *Journal of Object-oriented Programming*, **2**, (**4**), 39–43.

Goldberg A. and Robson, D. (1983) *Smalltalk-80: The language and its implementation.* Addison-Wesley, Reading, MA.

Hendersen-Sellers, B. and Edwards, J. M. (1990) Object-oriented systems life cycle. *Communications of the ACM*, (1990) **33**(9), 143–59.

Hewitt, C. (1977) Viewing control structures as patterns of passing messages. *Artificial Intelligence*, **8**(3), 323–64.

Minsky, M. (1975) A framework for representing knowledge. In *The Psychology of Computer Vision*, (ed. P. Winston), McGraw-Hill, New York, NY.

Parnas, D. L, Clements, P. C. and Weiss, D. M. (1985) The modular structure of complex systems. *IEEE Transactions on Software Engineering*, **SE-11**(3), 259–66.

Pinson, L. J. and Wiener, R. S. (1991) *Objective-C.* Addison-Wesley, Reading, MA.

Tanik, M. M. and Yeh, R. T. (1989) Rapid prototyping in software development. *IEEE Computer*, **22**(5), 9–10.

Taylor, D. (1990) *Object-oriented technology: A manager's guide.* Servio Corp., Alameda, CA.

Tesler, L. (1986) Programming experiences. *BYTE*, August.

Tracz, W. (1987) Confessions of a used-program salesman – fringe benefits. In 'The open channel', *Computer*, May, 109.

Tsichritzis, D. C. and Nierstrasz, O. M. (1989) Directions in Object-oriented research. In *Object-oriented concepts, databases, and applications*, (eds. W. Kim and F. H. Lochovsky), ACM Press, 523–536.

Wilcyzynski, D. (1988) A Common Device Control Architecture. The Savoi Actor, Proceedings of Autofact '88, November.

Winblad, A., Edwards, S. D. and King, D. R. (1990) *Object-oriented software.* Addison-Wesley, Reading, MA.

Yonezawa, A. (ed.) (1990) *ABCL: An object-oriented concurrent system.* MIT Press, Cambridge, MA.

Yourdon, E. (1990) *Modern structured analysis.* Prentice-Hall, Englewood Cliffs, NJ.

3 Object-oriented software: Relevance to manufacturing

S. ADIGA

3.1 Introduction

One of the most discussed topics in manufacturing in the 1980s was computer-integrated manufacturing (CIM). CIM, as the name suggests, is an effort to integrate activities in manufacturing through the medium of computers. It is now widely accepted that CIM leads to widespread improvement in productivity. Long-term benefits of CIM are based on the realization that CIM is neither hardware nor software for automation, but a strategy. It is a strategy related to computer-based sharing of resources and the availability of timely and appropriate information to optimize performance of the entire manufacturing organization.

Much research has been done on various aspects of CIM. Yet, today, there are still too many open issues as regards an architecture for CIM. These are related to hierarchical control, interfaces, communication standards, definition of subsystems etc. According to McLean (1987), it is questionable whether a consensus will ever be reached on these issues. Irrespective of the individual approach taken to the achieving of CIM, the topics of software development, communication and integration are considered crucial.

While we have not seen a generic framework for CIM applications gain acceptance in the industry, many specific CIM implementations have been successful. Weston *et al.* (1988) summarized the major limitations of specifically designed CIM systems as follows:

1. The need to 'reinvent the wheel' for the wide variety of manufacturing systems and industry requirements that are encountered.
2. The inability to reconfigure the system when previously unforeseen product and/or process changes are required.
3. The difficulties of achieving inter-company integration where information and functionality are shared across company boundaries.

We believe that the OO approach has the potential to offer generalizable solutions to overcome all the above limitations. The material in this section will cover mainly software development, communication and integration aspects of CIM. These broad topics address the issues raised by Weston *et al.*, among others.

3.2 Software development aspects of CIM

This section deals with some major issues related to software aspects of CIM, and how OO concepts may be used to address these issues.

3.2.1 Complexity and variety

Manufacturing systems are quite complex and varied in nature. It is impractical to imagine that a single solution or software package will address all the needs of all manufacturing firms. In an interview, John Clancy (Barrett, 1986), the Chairman of McDonnel Douglas, said that each manufacturing company's floor is unique, and that would have to be taken into account when creating a CIM system. Therefore, a practical approach to designing software for managing CIM systems is to build generic solutions to the greatest possible extent, and then to customize them to suit the needs of each firm. Thus, generic software object class libraries, customizable through subclassing, provide a good starting point in the design of practical software.

3.2.2 Abstraction issues

From manufacturing world to computer world

Manufacturing-related people think of their systems in terms of parts, conveyors, lathes, or drilling machines etc. In other words, they think in terms of 'objects'. An OO approach allows designers and programmers to construct software counterparts of manufacturing entities easily with little conceptual mismatch. As reported by Najmi and Lozinski (1989), this makes it easier for system designers to discuss models (underlying the programs) with users.

Aggregation problems

Not all CIM applications require the same level of detail in the description of the system state. For example, a production planning module (for wafer fabrication) in a microelectronics factory probably only needs to view the factory in terms of areas (vapor deposition, photolithography, ion implantation, diffusion etc.), while a short interval scheduling application will require a description of where the photolithography area is further expanded in terms of, say, coat, expose, develop and etch as shown in Figure 3.1. A real-time material handling system may require finer details on movement of wafers as shown in the third level.

Representation of important aggregations for different decisions being made in a factory is a critical feature of good software design in manufacturing. The ability to treat objects as a collection of other objects, and the inheritance feature of OO systems are useful in representing aggregations.

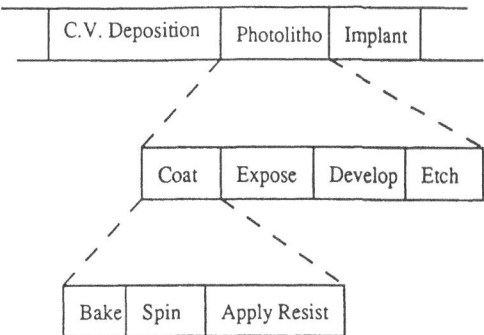

Figure 3.1 *Aggregations in representation of steps in wafer processing.*

3.2.3 Data management

Databases play a very important role in storing and retrieving data needed in manufacturing. Traditionally, three data models have been used in commercial databases: hierarchical, network, and relational. Object-oriented databases have only recently entered the field.

Because of the presence of hierarchies in manufacturing systems, i.e. hierarchies of parts-structure, routes, organization of people in management etc., a hierarchical database may appear as a natural choice. But there is also much data related to the interrelationship between entities that is not of a hierarchical nature. Moreover, problems of hierarchical databases in representing many-to-many relationships, and others, are well documented (Date, 1989).

Currently, relational databases are being used extensively in engineering applications. However, computer-aided design (CAD), process planning, and testing applications demand data types involving geometric forms, hierarchical data, and set matrices among others. Stonebraker *et al.* (1983) have argued for data types of a procedural nature. These heterogeneous demands suggest that an object-oriented database, which can handle complex data types conveniently, is more appropriate for engineering applications than a relational database. The relational model is relatively limited in the data types it can support. Databases are discussed in greater detail in Chapter 6.

3.2.4 Hierarchical control

Factory control systems (both manual and computerized) have distinct decision-making levels. The number and definitions of levels varies from situation to situation (Rhodes, 1987). A popular model consists of the following levels:

- corporate
- factory
- area (shop or department)
- cell (a collection of workstations)
- workstation (a collection of machines or devices)
- device

In a typical hierarchical model of control, control commands and decisions flow downward, and production status and performance information flow upward. The factory representation is therefore required to operate at different levels depending on the control or decision logic being implemented. As a result, different computer programs or plans are needed to reflect the respective needs, often at the cost of information to other levels. An OO approach allows data and procedures to be represented at different levels of abstraction as appropriate to the situation. Thus it is possible to present multiple views of the situation in a consistent and integral manner.

Further discussion on hierarchical control can be found in Chapter 4.

3.2.5 Simulation and control

Discrete event simulation has emerged as a powerful and popular tool for the analysis and design of manufacturing systems in the 1980s. Young *et al.* (1988) show that simulation can be used to support all stages of development of CIM systems, from specification to implementation.

Both simulation and control systems require a model of the real world. Ideally, one would like to be able to share for simulation purposes the state model developed for control purposes (or vice-versa), including both the structure and the data (for example, to simulate the future activity of the factory, starting from its current state). Also, since simulation is quite popular as a tool used to validate control strategies, control modules implementing these strategies have to be developed. Again, sharing these modules between the simulation and the control tool can improve productivity and also the consistency of the applications.

Since OO systems represent objects that correspond to real- life entities, and they are interchangeable since they conform to the same message protocol, we have a unique opportunity to develop a system that can be used initially as a simulation tool and, later, as a production or control software. Specifics of an architecture that makes this possible are presented in the next chapter.

3.2.6 Incremental development of CIM systems

The technologies, finance or experience required to build and install CIM systems may not be present in all companies. This has led many people to believe that the most appropriate way to implement advanced manufacturing technology is in an incremental manner (Bessant, 1985; Wittry, 1987). Prototyping reduces the risk involved in implementing large projects. Incremental development and implementation is possible with an OO approach as discussed in Chapter 2.

3.2.7 Customization and maintenance

Pan, Tenenbaum, and Glicksman (1989) identify two major shortcomings of the current CIM systems: they are difficult to customize and maintain; and they have very limited problem-solving and decision-making capabilities. The first short-coming can be addressed easily in an OOS system. Individual objects can be customized through the subclassing enabled by the inheritance feature of OOS. For example, if the application requires a representation of a lathe machine that is different from the one in the library supplied, a subclass Lathe can be created – say, NewLathe – that inherits all the functionality of Lathe with additional methods to enhance its functionality. Similarly, an undesirable feature can be overridden through a reimplementation in the new subclass. It has been shown elsewhere (Cox, 1986) that OOS are much easier to maintain than conventional software systems.

The kinds of decisions made by software is dependent on the methods used by the problem solvers. The operations research (OR) community has been working on mathematical models to support decision-making for years. More recently, rule-based systems based on ideas developed in artificial intelligence (AI) are rising in popularity because of their promise to capture the experience base of decision-making. Hybrid approaches are discussed by Kusiak (1988), among others.

The simpler flow of control and modularity offered by OOS can be used to enhance the flexibility of the decision-making capabilities of software using a hybrid approach (Adiga and Lin, 1990; Lefrancois and Montreuil, 1990).

3.2.8 Commercial CIM software

Software has become an important component of the automation activity. Automation equipment manufacturers as well as users are realizing that the functionality of today's automation equipment is determined, to a large extent, by the software contained in the system. According to Dietrich Ernst (1991), software costs today are often higher than the development costs for the corresponding hardware.

According to one estimate (appearing in 'Electronic Business'), associated hardware, software and service costs can put the commercial factory floor management systems in the price range of $500,000 (Burrows, 1991). Manufacturers have an option to buy or license software packages from software suppliers instead of developing their own software. When you buy a vendor's software, your views of conducting business often have to match the vendor's own views on the subject. It is said that this option usually meets only a small part of all software needs and the customers will end up modifying or enhancing the purchased software. Therefore, it is almost inevitable that entry into the CIM area will involve a good amount of in-house software development.

Appleton (1986), an industry leader in CIM software issues, mentions that even when a firm buys commercial software sold as suites of 'integrated' software modules, these suites often cannot be effectively integrated with one another. He further adds that software modules are often written in-house to perform manufacturing planning and control functions beyond those being performed by implemented modules already purchased. We believe that the sharing of common classes, and the ability to reuse these classes by the successive specialization required by different applications, will encourage better integration of software.

OO technology promotes the idea of building software systems by putting classes of objects together. The concept of hardware silicon chips in integrating electronic circuits has changed forever the way hardware engineers build systems. OOP allows a similar packaging and reusable effort in software development. Cox (1986) describes a Software-IC as a package of programming effort that is independent of the specific job at hand and highly reusable in future jobs. Following the Software-IC concept, the programming effort could become one of selecting the right modules (or objects) and tying them together with messages to reflect the configuration required. Thus, the relationship between system developers and engineers turns into one of suppliers and consumers of Software-ICs.

3.3 Communication issues

Electronic data communication is an important part of modern manufacturing system. Communications among various computers and control devices remain problem areas in the implementation of CIM systems. The development of the standards MAP (Manufacturing Automation Protocol) and OSI (Open Systems Interconnection) contributes to solutions in this area. We confine our discussion here to some interesting issues relating to the application of the OO paradigm for data communication in manufacturing.

Objects provide a loosely coupled way of communication. The basics of inter-object communications were discussed in Chapter 2. In this chapter, we discuss issues that arise due to inter-object communication on remote machines.

3.3.1 Inter-object communication on remote machines

A common approach to handling inter-process communication (over remote machines) is to put data to be communicated into packets and transmit the packet using some network protocol that is common across machine architectures. The approaches to doing this in OO systems include the following:

1. The passing of references to other objects as parameters. This leads to problems when references, which are usually addresses, are dependent on the local machine and make no sense to the remote machine. A solution to this problem is to have address-independent identifiers for objects. When messages are sent to objects, they can be directed to the object at a relevant location. Lozinski (1991) discusses an implementation of this scheme.
2. In Remote Procedure Call systems and some other implementations, the passing of parameters by value for remote invocations. This, as pointed out by

Hutchinson and Walpole (1991) changes the semantics of the object model for remote operations and destroys distribution transparency.

3. A 'call-by-proxy' approach (Hutchison and Walpole, 1991). The previous approach obviously may introduce consistency problems. This solution is more general: it calls for the creation of a proxy object that is a representative of a remote object on a local machine. Messages to the proxy object are forwarded to its remote object.

4. The use of distributed objects, i.e. instances that are shared across machine address spaces. The best way to accomplish this is still an active research area; we are just beginning to see commercial products implementing this idea.

5. Distributing a copy of an instance object. A software implementation of this approach is discussed by Knolle *et al.* (1990).

The selection of an appropriate method for an application depends on the size of objects being moved around, the nature and frequency of operations to be performed, the costs of moving the object, etc.

Decouchant's (1989) paper describing the design of a prototype distributed object manager discusses many interesting problems of, and possible solutions to, the managing of objects over network nodes.

3.3.2 Fault tolerance

Some aspects of object-to-object communication may lead to a fault-tolerant design of a software system. In an asynchronous communication mode, the decision to wait for a reply or not is made by the system designer or programmer. Fault tolerance is implemented easily, as the object requesting service is not blocked when the service requested is being processed by the object providing the service.

If the network breaks down while the local cell is controlled by OO programs, information on the part is still available to continue operation. All the instantiated objects carry the most recent data in their instance variables. Such a facility promotes fault tolerance at the cell level.

Objects have self-awareness. That is, you can verify if an object supports a particular method (needed to respond to a message) by asking it before sending it the message. This allows potential runtime errors to be avoided. In some languages such as Objective-C, this checking can be done in runtime; in others such as C++, it is done in compile time. This feature allows one to avoid or anticipate faults in programs resulting from objects receiving unknown messages.

3.4 Integration problems

A key question to be answered concerning the integration problem today is: how do we integrate several subsystems, each with its own controllers and computers, in a multi-vendor environment? If the subsystems are tightly bound, it is very difficult to change individual systems when better technology, process or customer demands make such a change desirable.

Empirical evidence suggests that one does not achieve integration by just wiring such islands together in some fashion; it has to be done as a part of an overall plan (Shunk and Filley, 1986; Thomson and Graefe, 1989; Kellso, 1988). A popular approach often proposed during the 1980s is the use of a common database (usually relational) to store data for integrating manufacturing systems; this approach will be analyzed next.

3.4.1 The common database approach

Historically, CIM applications for different purposes (for example, process control, production scheduling etc.) have been developed independently. As these applications have at least partially common data needs, this practice has led to data redundancy and, consequently, data inconsistency. A separation of the application logic and the data (grouped into files) paved the way for databases, where the data needed for different applications is stored in one place. The advantages of this architecture (a central database shared by application programs) are well-known (Date, 1986): i.e. the reduction of redundant data, avoidance of inconsistency, data sharing, standard enforcement, integrity maintenance, etc. The leading commercial CIM systems for semiconductor fabrication (e.g. WORKSTREAM, PROMIS) adopt this approach, in part at least.

The database provides the common platform or model through which the different applications, present and future, have their data needs fulfilled. But designing that model in a way to accommodate the different views required for all individual applications, while allowing room for future growth, is an extremely challenging task. In particular, if all data are treated in the same way, this model could turn out to be a gigantic one. It could seriously affect the response time if all queries are to be handled by one database.

This problem is encountered with some commercially available CIM systems, where, for example, Work-In-Progress (WIP) extracts may be obtained only in batch mode a few times a day, as they require prohibitive time for real-time access. In fact, not all applications (and therefore not all data) have the same degree of urgency: while a process engineer may accept a five-minute wait for the result of some yield analysis, the operator requesting advice from the scheduling system expects an immediate answer. Incidentally, usually the boundary between urgent and 'not-so-urgent' data corresponds to the boundary between current and historical data. Keeping the current image in RAM (Random Access Memory) while having historical data on disk would be one way of handling that problem.

Though it is applicable to most situations, this approach may not solve all problems as some specific applications such as Statistical Process Control (SPC) have unique requirements. SPC requires measurement data from previous runs to detect undesirables trend in equipment performance: resulting warnings, if any, need to be issued at once. These requirements must be considered individually.

In addition to the points discussed above, the conventional database approach suffers from other limitations. It does not readily meet the need for flexibility required by future applications development. For example, all current applications may need to see the workstation capacity as an average value. When developing the database, it is therefore natural to have a workstation table (i.e. relation) with, for example, an 'average capacity' field. The applications will know of this structure and, when needed, will direct their query about capacity data toward the workstation table. When, however, a new application requiring a finer level of detail (say, distinguishing between different causes of downtime) is introduced, the workstation table must be expanded to accommodate the needs of this new application (by introducing new fields corresponding to this new data).

Now, the database administrator is left with three choices: change every existing application to consider the new database structure (a clearly impractical solution), keep capacity data both in an aggregate and in a detailed form (leading to data redundancy and potential inconsistency), or use such mechanisms as 'view' in SQL (Date, 1986). However, even in the latter case, a different name must be used for the view and the base table, which may result in modifications of the existing applications. Furthermore, the two data structures may coexist if the detailed information is required for some, but not all equipment. Then the old application programs would have to know whether to direct their query toward a base table (in the case of equipment for which the simplified data structure is still in place) or a view (in the case of equipment for which the new, more detailed data structure is now available).

The object-oriented paradigm provides an alternative, more satisfying answer to these problems. The real world can be modeled as a set of software objects corresponding closely to their physical counterparts: lots, machines, workers, etc. The instance variables of these objects maintain the data describing their current state, and can be accessed by the application programs by means of messages. This mechanism provides an effective shield for the data (through data encapsulation) and addresses the concerns expressed above.

A change in the internal data structure of the object need not affect its interface with the application programs. Considering the previous example, a workstation object needs to respond to the message 'averageCapacity'. In the first case, where this average capacity is stored internally in the workstation object, responding to the message merely involves returning the value of the relevant instance variable. In the second case, where average capacity is the result of a computation involving more detailed data, the receipt of the message will trigger this internal computation. The advantage is two-fold: first, there is no need for modification of the application program, since its interface with the object (the message 'averageCapacity') is unchanged, and second, whether the workstation object actually maintains a simplified or a detailed data structure is transparent to the application program, and both types of objects may coexist without the program being aware of it (polymorphism).

An OO view of the systems, complete with data encapsulation and availability of messages to synchronize cooperating processes, can provide a robust implementation that is not affected by design changes. We are not suggesting that making subsystems or 'islands of automation' talk to each other by sending messages is the way to integrate them. We propose that an object model of the enterprise be used as the reference for integration. It is a fact that planning at the system level is an important part of the design of an object model. Such an approach will force system designers to consider interactions of subsystems before designing them.

3.4.2 A common object model

The use of a database for integration purposes was the first step in the right direction: that is, common data moving around the factory, connecting different functions or departments. Since data is raw and needs to be interpreted in different ways, perhaps, unique to each manufacturing function, this approach results in much traffic in data and problems as described earlier.

As shown in Figure 3.2, data, information and knowledge can be seen as using different levels of abstractions of the real world. Since an object encapsulates the data and procedures that provide the context in which they are used, an object model is basically an information model of the system. Use of an object model as the central unifying theme for CIM is a natural evolution of the earlier approach of using a database as the integrating influence.

Designing CIM systems to an information model of the factory (in an OO representation) will provide the necessary conceptual, as well as physical, binding between the subsystems comprising each CIM system. This approach makes integration possible while allowing data to be stored wherever it is convenient to do so. Communication through messages provides the loose coupling between subsystems required in a dynamic environment. Implementation of the design in

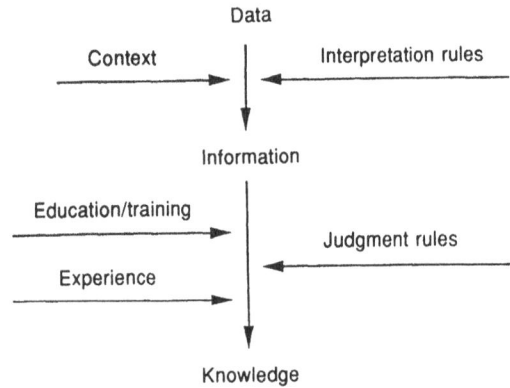

Figure 3.2 *Data, information and knowledge.*

an OOP language minimizes the conceptual mismatch between design and implementation stages.

It may appear attractive to use knowledge as the integrating influence in CIM. The technology made possible by research in artificial intelligence through expert systems (Ignizio, 1991) enables a scheme to be proposed which uses a knowledge base as the central entity. As shown in Figure 3.2, knowledge is a subjective construct. We are wary of basing the design of a factory-wide system on subjective knowledge that is expected to change through further experience. Ham and Kumara (1990) correctly point out that no two people may have a common interpretation of the concepts described in a knowledge base. Moreover, knowledge acquisition is a known bottleneck in building such systems. While we do acknowledge the role of knowledge-based systems in solving difficult problems in manufacturing, we do not see them as an integration mechanism.

An object model is a collection of objects (and their hierarchies) with definitions of instance variables and methods. Adherence to a common object model provides the users with some technical standards for intermodule communication. Object interfaces form the basis for these standards. Standards provide the glue that binds the components together. Adherence to a common model helps people plan their applications or individual solutions based on common resources captured in the object model.

3.5 Conclusion

Problems in developing software for manufacturing systems have been discussed in this chapter. While there are many advantages in the OO approach, we discuss only issues related to software development and integration, and how best they can be addressed by the OO technology.

The OO approach based on communication via messages provides the means for flexible integration where individual applications can remain distinct but can interact in a reconfigurable manner as desired by the user. One may build systems that can be upgraded or enhanced in an incremental manner. Systems can be designed with minimum mismatch between the conceptual and physical aspects of design and implementation. This potential of the OO approach may not be realized to the fullest extent possible without an overall framework to build such systems. In Part Two, we propose such a framework or system architecture for this purpose.

References

Adiga, S. and Lin, W-T. (1990) An object-oriented architecture for production scheduling systems. Technical Report No. ESRC 90–1, Engineering Systems Research Center, University of California, Berkeley, CA.

Appleton, D. S. (1986) Integration technology. *CIM Technology*, Spring, 16–18.

Barrett, C. (1986) An advocate's view of CIM. Interview with McDonnell Douglas' John Clancy. *Computer Graphics World*, May, 81–86.

Bessant, J. (1985) The integration barrier: Problems in the inplementation of advanced manufacturing technology, *Robotica*, April–June, 97–103.

Burrows, P. (1991) Factory management systems bring planning to the plant floor. *Electronic Business*, February 4, 63–65.

Cox, B. J. (1986) *Object-oriented programming*. Addison-Wesley, Reading, MA.

Date, C. J. (1990) *Introduction to Database Systems*, 5th edn., Addison-Wesley, Reading, MA.

Ernst, D. (1991) The forces propelling automation technologies. Interview in *Siemens Review*, January, 4–7.

Glassey, R. and Adiga S. (1989) Design of a software object library for simulation of semiconductor manufacturing systems. *Journal of Object-oriented Programming*, **2**(4), 39–43.

Ham, I. and Kumara, S. R. T. (1990) Intelligent computer integrated manufacturing (I-CIM): Research perspectives. *Proceedings of IAIS90*, 7–16.

Hutchison, D and Walpole, J. (1991) Distributed systems and objects. In *Object-Oriented Languages, Systems and Applications* (eds. G. Blair, J. Gallaghar, D. Hutchison and D. Shepherd), Halstead Press, UK, 223–43.

Ignizio, J. P. (1991) *Introduction to expert systems*. McGraw-Hill.

Kellso, J. (1988) *Implementing Integrated Solutions on the Factory Floor*. Presentation CIM-IE Workshop, Stanford University, Palo Atto, CA, August.

Knolle, N., Fong, M. W. and Lang, R. W. (1990) SITMAP: A command and control application. In *Applications of Object-Oriented Programming* (eds. L. J. Pinson and R. S. Wiener), Addison-Wesley, Reading, MA, 28–65.

Kusiak, A. (1988) EXGT-S: A knowledge-based system for group technology. *International Journal of Production Research*, **26**(5), 887–904.

LeFrancois, P. and Montreuil, B. (1990) An object-oriented knowledge representation for intelligent control of a workstation in a rolling-mill facility. Technical Report 90–51. University Laval, Canada.

Lozinski, C. (1991) Why I need Objective-C. *Journal of Object-Oriented Programming*, September, **4**(5), 21–8.

McLean, C. R. (1987) Interface concepts for plug-compatible production management systems. *Computers in Industry*, Elsevier Science Publishers B. V. 307–18.

Najmi, A. and Lozinski, C. (1989a) Managing factory productivity using object-oriented simulation for setting shift production targets in VLSI manufacturing. *Proceedings of Autofact '89 Conference*, 3:1–3:14.

Pan, J. Y.-C., Tenenbaum, J. M. and Glicksman, J. (1989) A framework for knowledge-based computer-integrated manufacturing. *IEEE Transactions on Semiconductor Manufacturing*, **2**(2), 33–46.

Rhodes, J. S. (1987) Integration of the automated factory. Technical paper presented at *AMS'87, Advanced Manufacturing Systems Exposition and Conference*, Chicago, IL.

Shunk, D. L., and Filley, R. D. (1986) Systems integration's challenges demand a new breed of industrial engineer. *Industrial Engineering*, May, 65–7.

Stonebraker, M., Rubenstein, B. and Guttman, A. (1983) Application of abstract data types and abstract indices to CAD data bases. *Database Week*: Engineering Design Applications.

Tessler, L. (1986) Programming experiences. *BYTE*, August.

Thomson, V. and Graefe, U. (1989) CIM – A manufacturing paradigm, *International Journal of Computer Integrated Manufacturing*, **2**(5), 290–7.

Weston, R. H., Gascone, J. D., Rui, A., Hodgson, A., Sumpter, C. M. and Coutts, I. (1988) Steps towards information integration in manufacturing. *International Journal of Computer Integrated Manufacturing*, **1**(3), 140–153.

Young, R. E., Vesterager, J., Wichmann, K. E. and Heide, J. (1988) Simulation uses in CIM development, *Journal of Computer Integrated Manufacturing*, **1**(1), 50–4.

Summary: Part One

S. ADIGA

Object-orientation: General

Encapsulation of data and methods within an object, and communication by exchanging messages are the most fundamental attributes of the object-oriented paradigm. Objects in the software world can be designed and named to match the physical counterparts in the external world. This reduces the conceptual mismatch between the programmers' and actual world views. The concept of inheritance reduces the amount of redundant code in the system. Other properties, such as inheritance, promote the sharing of code and the propagation of change among objects. Dynamic binding and polymorphism enable more understandable and flexible design of software.

Perhaps the biggest contribution comes from the realization that OOP is not a programming method but a software packaging technology. With the right library of Software-ICs, programming for manufacturing systems might become just the task of choosing and interconnecting objects of machines, operators, parts etc., with the necessary measurement and analysis type of objects.

Object-orientation: Manufacturing systems

While there are numerous advantages in applying object-orientation to CIM-application development, we mainly focus on the issues related to software development and integration.

Those who design manufacturing systems have realized by now that there is no one-size-fits-all solution to problems in manufacturing. Building custom applications off a common repository of software objects is an attractive idea, as it helps to achieve a conceptual integration.

Need for a framework

From the above arguments, it may be seen that the OO paradigm has considerable potential to improve the software development process for CIM projects. But this potential may not be realized to the fullest extent possible without an overall framework to guide the development of such projects. We recommend that the design of OO software in an organization should involve the following steps:

1. The visualizing of an overall architecture for all the software components contributing to the solution of a particular problem or the system being built. For example, if the current software development job relates to the process planning function, an overall architecture will highlight the major modules in this system and the ways in which they relate to each other, and how the system as a whole relates to other software components in the organization such as product database, design database, etc. The software architecture should address the questions of which software libraries are required and what the interfaces might be. In addition to the determination of external interfaces, this architecture would help in the setting of boundaries to the libraries being designed.

   A reference model that can serve as the central design framework for an object-oriented software development effort in manufacturing organizations is discussed in Part Two.

2. Once it is known what libraries are needed, the next level of design involves answering the question: what objects should be included in the library? A method for identifying the objects and required behavior in the OO software to be built is discussed in Chapter 4.

3. The specifications for the objects need to be implemented in a programming language. The list of programming languages/systems claiming to be object-oriented is growing every day. Two popular languages currently being used for implementing manufacturing applications are Objective-C and C++. Therefore, an in-depth comparison of these two languages is provided by Milind Gadre in Chapter 7. The discussion is supplemented by some references to other options in the field.

For further study

Biggerstaff, I. 1989. *Software reusability Volumes I and II*. ACM Press, Addison-Wesley.

Booch, G. (1991) *Object oriented design with applications*. The Benjamin/Cummings Publishing Co., Redwood City, CA.

Booch, G. (1986) Object-oriented development. *IEEE Transactions on Software Engineering*, **12**(2), 211–221.

Brooks, F. P. (1987) No silver bullet: Essence and accidents of software engineering. *IEEE Computer*, **20**(4), 10–18.

Buzzard G. D. and Mudge, T. N. (1985) Object-based computing and the ADA programming language. *IEEE Computer*, March, 11–19.

Cox, B. J. (1984) Message/object programming: An evolutionary change in programming technology. *IEEE Software*, **1**(1), 50–61.

Cox, B. J. (1990) Planning the software revolution: The impact of object-oriented technologies. *IEEE Software*, November, 25–33.

Dean, H. (1991) Object-oriented design using message flow decomposition. *Journal of Object-Oriented Programming*, **4**(2), 21–31.

Decouchant, D. (1989) A distributed object manager for the Smalltalk-80. In *Object-oriented concepts, databases, and applications*, (eds. W. Kim and F. H. Lochovsky), A–W Press, Addison-Wesley, 487–522.

De Marco, T. (1978) *Structured Analysis and System Specification*. Yourdon, Inc.

Peterson, G. E. (ed.) (1987) *Tutorial: Object-oriented computing, Volumes I and II*, IEEE Computer Society Press, Washington DC.

Ramamoorthy, C. V. and Sheu, P. C. (1988) Object-oriented systems. *IEEE Expert*, Fall, 9–15.

Stroustrap, B. (1986) *The C++ reference manual*, Addison-Wesley, Reading, MA.

Taylor, D. (1990) *Object-oriented technology: A manager's guide*. Servio Corporation, Alameda, CA.

Part Two
Design and Implementation Techniques

4 Towards an object-oriented architecture for CIM systems

S. ADIGA and P. COGEZ
University of California at Berkely, USA

4.1 Introduction

An architecture for a system helps one create an integrated vision. A computer architecture is composed of two elements: a structure and an organization (Lent, 1991). Structure describes the components and their interconnections. Organization presents the dynamical appearance of components and their management according to the selected operational principles. In this chapter, we propose an object-oriented system architecture for CIM applications. We identify a generic framework (i.e. a basic structure – Ehrlich et al., 1980) for the architecture and describe necessary interconnections. We also provide implementation perspectives based on OO concepts.

Designing a (software) framework is like developing a theory (Wirfs-Brock and Johnson, 1990). The theory in this context may be tested by building a prototype. It is validated through industrial implementations by reusing the framework. This architecture is being prototyped through extensions to BLOCS (Berkely Library of Objects for Control and Simulation) in the Engineering Systems Research Center of the University of California at Berkeley, USA.

4.2 A conceptual framework

It is a fact that specific problems involved in applications such as process planning, production scheduling or controlling process variations etc. require specialized solutions. In the past, these solutions have led to the evolution of systems that formed 'islands of automation'. In Chapter 3, we argued that an integration effort is best served by adopting a common philosophy and a common OO model for building these different applications. We would like to add that a common structure will improve this effort. The need for this is supported by industry's experience in recent years that a lack of clear-cut boundaries among different elements of a system hampers growth and prevents reusability of data between different applications. Additionally, it makes system maintenance more cumbersome.

The task of designing a generic framework is made more difficult by the fact that commonly accepted guidelines or specifications for factory automation modules do not exist (McLean, 1987), Defining a framework is similar to designing a system model. A key issue in the design of system models is that early design decisions condition the ability of future applications to model the physical system adequately, and affect growth in functionality. It is therefore essential that this design should be at a sufficiently high level and open, as it is impossible to forecast at the outset all the needs of future applications.

Our framework is based on a specific modularization, or partitioning, to serve as a skeleton for building most CIM systems. The modules have been defined at a high level to promote reusability and interchangeability. They also reflect the basic system functions that we see in most computer system applications in manufacturing. We believe that this framework will allow us to take advantage of OO concepts in an elegant and yet practical manner.

The primary concern of a manufacturing firm is to transform inputs in the form of raw materials into finished products to be sold to customers (to satisfy some need). For decades we have been devising information systems, heuristics and algorithms to make decisions in planning, monitoring and controlling activities as part of this transformation process in manufacturing plants. Therefore, most manufacturing systems can be viewed as consisting of two parts: decision and plant, or physical systems, as described in Figure 4.1.

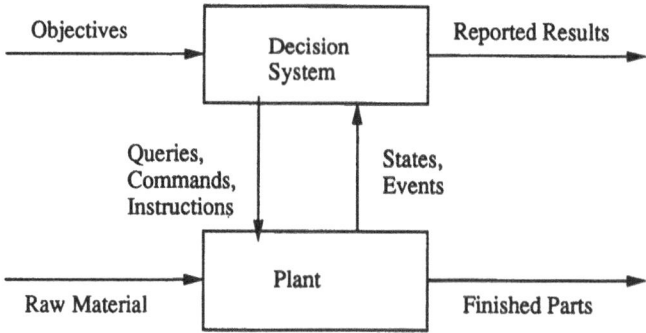

Figure 4.1 *A high level view of a manufacturing system.*

Decision systems include not only automated systems (analog and digital) but also manual systems. Plant may consist of human operators, or machines with local or embedded controllers such as numerically controlled (NC) machines, used in the transformation process. The communication from the decision system to the plant may be in the form of queries, or commands/instructions. Commands are meant for automated devices whereas instructions are issued for humans. The plant or physical system communicates its state or events that take place, as required, to the decision system. The decision system may also have a hierarchical structure as discussed in the previous chapter.

We can further decompose the overview of Figure 4.1 into specialized components as applicable to computerized application systems. Though most of the computerized applications in manufacturing have specific decision requirements, we can identify a typical decision system as a combination of the following modules (or subsystems).

1. A computer model of the manufacturing plant: Most applications are developed with a particular view of the plant in mind, i.e. they need a reference view of the plant. This model reflects the state(s) of the manufacturing plant or the physical world supplying the context for the decisions to be made. Since it is an abstraction of the state of the system, we will call this the state model.
2. Decision or control module: This module contains the decision or control logic used in solving problems in the application. The control logic is expected to work on the state model to achieve the desired action or results. The idea of separating plant from control has been advocated in work on control theory and practised by mechanical engineers for years. This module supports the decision process. Therefore, it may include monitoring, planning or control functions. Some examples in this chapter refer to control applications and may refer to this module as the control module.
3. An interface between the state model and the plant or physical world: This module allows one to define the state model and acquire data from the physical world.
4. An interface between the decision module and the plant (or the physical world): This module has the responsibility of receiving the requests from users or the real world in general, and returning the system recommendations or decisions.

Figure 4.2 shows the interrelationships between modules that make up a typical decision system and the physical system.

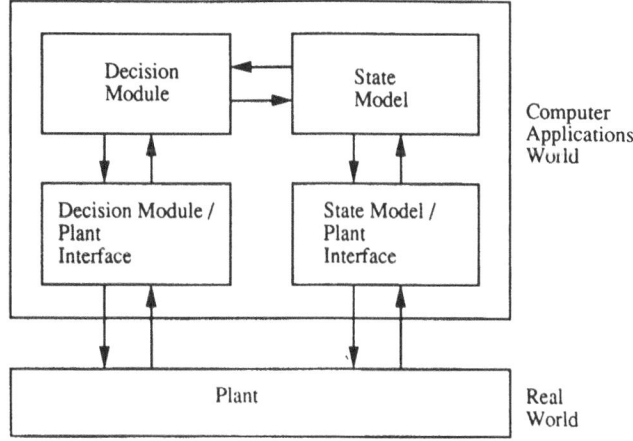

Figure 4.2 *A general framework for computer application systems in manufacturing.*

Details of the modules are discussed next. We use the term 'module' to mean a subsystem or an identifiable component of a system. A system is a combination of elements intended to act together to accomplish an objective (Palm III, 1986).

4.3 Description of the modules

4.3.1 State model module

We propose to use an OO model as the basis for deriving the common state model to be shared between applications. This model, when instantiated with specific state information, will also supply the necessary data for running some applications. We have already discussed the advantages of this approach over a common database approach driven by a relational data model in Chapter 3. This OO representation brings all the benefits previously discussed.

An instantiated object model allows active interaction between that model and the application programs. However, we have seen that besides accessing the state model, the application programs need to be informed of the changes triggered by the decisions of other programs or, more generally, of any relevant changes occurring in the plant. For example, if a measurement result has been entered in the database through the real world/state model interface, the statistical process control (SPC) application should be informed to carry on analysis of the equipment behaviour and issue warnings, if appropriate.

In the traditional database approach, these problems would call for the development of a 'shell' of programs around the database itself to detect the significant changes (in the example given above, writing to a measurement table) and relay the information to the relevant application program. Again, any change to the database structure would most likely require modifications to the above shell of programs. This makes the maintenance and future growth of the CIM system a painful task.

Objects in the state model need to do more than just reflect the state of the plant. An object receiving a message may have to send additional messages to inform the relevant application programs. This is to be contrasted with the more static role played by objects that are required only to update or report their states upon receiving a message.

4.3.2 Decision module

The use of the object-oriented paradigm in the modules implementing decision logic is not as clear-cut or obvious as it was for the state model. An OO approach is a natural one (compared to a non-OO approach) when developing decision modules to be applied against an object-oriented state model. For example, the decisions of a statistical process control module (such as when to issue a warning) may rely only on information at the current operation. In such a case, it would be easy to implement a decentralized decision-making algorithm relying on objects in the decision module that are mirror images of the operation objects living in the state model.

It is very likely that not all decision modules will require the same level of detail in the description of the system state. The advantage of using similar representations is more pronounced, perhaps, when different levels of aggregations are used. A production planning application will probably need to view the factory in terms of areas (photolithography, implant, diffusion), while a shop-floor automation application will require a description in terms of individual equipment. The state model should therefore be able to operate at different levels depending on the control logic being implemented. Multiple decision modules implementing control logic can be applied against a single state model that provides them with appropriate levels of aggregations and abstractions (as in Figure 4.3). This approach leads to an alternative way of handling multi-level problems to the traditional hierarchical control approach advocated in the body of manufacturing systems control literature.

4.3.3 Interface modules

The interface is an important part of a computer application as it may influence the success of the implementation. The interface between a state model and the physical world may be manual, automated, or a combination of both (semi-automated). An example of a manual interface is an operator keying in data at the start of an operation on a given machine for a specific lot. An automated interface is, for example, found in an assembly line where circuit boards are automatically inspected for faulty insertion of resistors. A mixed interface is, for example, an operator using a bar-code reading system to enter the fact that he/she has completed processing a part.

Irrespective of the quality of the control logic, a system that requires operators to spend half their time inputting data is unlikely to be successful on the factory floor. Likewise, a good control logic/physical world interface including explanation facilities is crucial, as many users tend to accept the system recommendations only if they feel that they understand them.

Another desirable characteristic of a CIM system is that different applications should share the same user interface. This has the potential effect of easing the burden on users. Of course, since these applications have different functionalities, the interface to different applications cannot be identical in every detail. Nevertheless, the interactive sessions with the user should, as far as possible, follow the same pattern. This can be achieved by a system of menus calling the application-specific parts of the interface, which in turn should follow some standard for screen presentation. A good example of this has been achieved in the personal computing area with the software packages written for the Macintosh computers, where a user familiar with one package can often get started on his/her own with another package. A common model supporting different applications allows a domain-specific user interface to be supported while conveniently maintaining some common traits of the system. A design based on a common model will help similar results to be achieved in manufacturing systems.

4.4 Communications between modules

To be able to further define the characteristics of the different modules of the proposed CIM architecture (in particular those of the shared state model), we need to examine the interaction between the modules. As a general rule, this interaction must be well defined so as to ensure 'plug-compatibility' of the control modules. Plug compatibility of modules implies that one should be able to add or remove modules without having to modify the shared model, or vice-versa. In other words, while most of the applications of the OO paradigm implement what Cox (1986) calls 'IC level' modularity (an analogy with integrated circuits in electronic hardware), our architecture is aimed at achieving a 'circuit board level' modularity.

Figure 4.3 illustrates the flow of information taking place between the different modules. The interfaces have been omitted to keep the diagram simple.

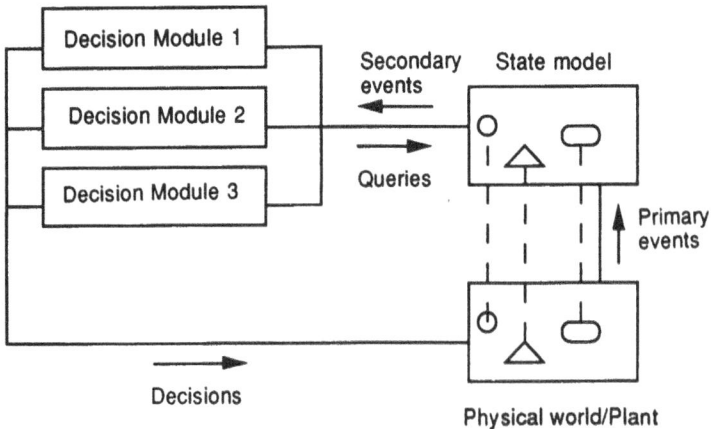

Note: — — — Signifies object-to-object correspondence

Figure 4.3 *Intermodule communication.*

4.4.1 Interactions between plant and state model

The state of the plant or physical system changes according to events (e.g. a machine failing, an operator bringing a lot into a cell, etc.). Events that are of significance for any one of the control modules should be passed on to the state model where they trigger appropriate changes in the values of the state variables.

Messages carrying the occurrence or termination of events are the main signals connecting the physical world and its counterpart, the state model.

4.4.2 Interactions between state model and decision module

Changes in the state model as a result of an event may, in turn, trigger the application of a control module. For example, a dispatching module is likely to be invoked, among others, every time a machine goes idle. Therefore, communication from the state model to control modules is also according to events.

To ensure maximum independence between the modules, the state model should not be required to make decisions that match information with the needs of respective decision logic or control modules receiving that information. One way of achieving this is to broadcast the events (resulting from changes) to the different modules, which will decide whether the changes are relevant for them.

Sometimes, decision modules may need additional state information to be able to carry out their respective decision logic. They obtain this information by sending queries to the state model.

This protocol can be easily applied to periodic decisions – as opposed to event-based decisions – by including current time as one of the state variables.

4.4.3 Interactions between decision module and plant

The outputs of a decision module directed towards the physical world are decisions or commands. These are implemented through the decision module/physical world interface. This may take the form of an operator reading an instruction on the screen and loading the corresponding machine or, in a fully automated environment, of a signal sent to numerically-controlled equipment.

4.4.4 An example

The following example illustrates the various interactions discussed above. An operator has just completed a production task on a given lot with a given machine. He records the event 'production task completed' in the computer system. This event is transmitted to the state model (i.e. a message is sent to the state model recording this event).

In the state model, copies of the old states of the lot and the machine objects are written to a historical database; the state of the machine object is changed from 'busy' to 'idle', the state of the lot is changed from 'in process' to 'waiting', and the location of the lot along the route is updated.

These secondary events are, in turn, broadcasted to the different decision modules. Among these, a release module implementing, say, a continuous fixed workload strategy would react to the fact that a lot has just completed a (process) step. It would check whether that step was at the bottleneck by sending queries to the state model. Likewise, the dispatching module will react to the fact that a machine is now idle by checking whether it is possible to load a new lot on it. If it is possible to do so, the respective operator will be prompted with a message

'Load lot'. The operator's acceptance of this message will result in the start of a new event to start the production task on the machine.

Data in the history database may be used later by specific control modules to determine machine utilization, perform yield analysis etc.

To some extent, the Berkeley Library of Objects for Control and Simulation of Manufacturing (BLOCS/M) conforms to the meta model discussed above. Objects such as workstations and lots make up the state model, and their instance variables are updated in response to events such as the start of a production task generated by simulation specific objects called future events. These events substitute for real world events in a factory. (In the current version of BLOCS, these future events also contain state information. The separation of that information – which, according to our framework, belongs to the state model – from the simulation-specific data is underway.)

To experiment with new control strategies, the researcher only needs to define control objects that obtain the required data by querying the state model. The decision is then translated into the creation of a new future event in the simulation module. When the simulation clock reaches the wake-up time of the future event, the corresponding event is generated and triggers appropriate changes in the state model.

4.5 Special concerns

Issues concerning data management in modern manufacturing and hierarchical control have received much attention in recent times. We now discuss these topics in relation to our architecture.

4.5.1 Data management issues

If all data were to be treated in a uniform manner and if all queries were to be handled by one database, the state model could turn out to be a gigantic database, its size seriously affecting the response time. This problem is encountered with some commercially available CIM systems, where, for example, WIP (Work–In–Progress) extracts may be obtained only in batch mode a few times a day, as the time required for real-time access is prohibitive. In fact, not all applications (and therefore not all data) have the same degree of urgency. While a process engineer may accept a five minute wait for the result of some yield analysis, the operator requesting advice from the scheduling system expects an immediate answer.

Keeping the current image in RAM (Random Access Memory or the semiconductor memory of the computer) while having historical data on disk is one way of handling the above problem. It means that the current state of the physical system is being maintained in RAM. The instance variables of these software objects maintain the data describing their current state, and they can be accessed by the application programs through messages. Thus the applications gain quick access to current data. Often, the boundary between urgent and not-so-urgent data corresponds to the boundary between current and historical data. Specific implementations of this conceptual scheme may vary according to the strategies used for achieving fault tolerance in applications.

Historical data, by definition, are not subject to change either in structure or in content (once they have been initially recorded, and barring rectification of mistakes). Therefore, they do not need this programming shell as provided by the objects, and could therefore be maintained in a relational database. Figure 4.4 is an illustration of a state model where the current image of the physical system is represented by objects, while historical data is managed by a traditional database system. An object-oriented database offers considerable advantages by reducing the conceptual mismatch between a software object and its representation in storage.

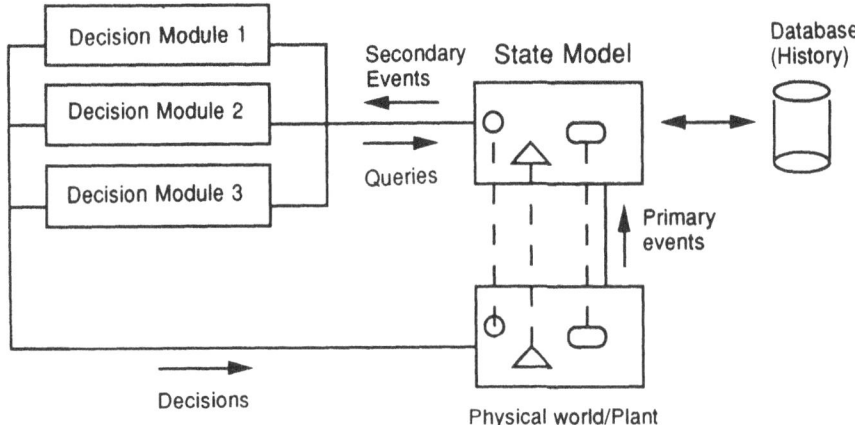

Note: — — — Signifies object-to-object correspondence

Figure 4.4 *State model and database connection.*

A general scheme containing the factory organizational hierarchy and timing requirements was discussed by Meyer *et al.* (1989). This scheme has been adapted in figure 4.5.

Our approach to data management may not solve every problem, as some specific applications such as statistical process control (SPC) have unique requirements. SPC requires measurement data from previous runs to detect any undesirable trend in equipment performance; any resulting warnings need to be issued at once. These requirements must be considered on an individual basis.

One of the assumptions (or requirements) of the above discussion is that, in a typical application built according to our architecture, the physical objects (lots, machines, workers, etc.) in the factory are expected to update their software counterparts in the state model upon changing their individual states.

	Time Horizon	Time Period	Decisions/Actions
enterprise	years	year	product mix
factory	year/ quarters	month	production plans
shop or area	month	day	work release, production schedule
work cell	day	minute	work transportation
work station	minute	second	dispatching/setup
machine/device	milli-second	milli-second	manufacturing operation (atomic level)

Figure 4.5 *Organizational hierarchy in a factory (after Meyer et al., 1989)*

4.5.2 Hierarchical control

Much research in manufacturing systems control in recent years has focused on hierarchical control. In this approach, commands are issued downward and status information is sent upward through layers of responsibility (Jones and McLean, 1986). This view fits well with the way most corporations are organized. Similarly, organizing computer systems in a hierarchy helps us manage its complexity. This hierarchical view is advocated in both the control and design of CIM systems (Malakooti, 1989).

In our architecture, the hierarchical nature of organizations is reflected in the hierarchical class structure of the objects in the state model. Control (decision) modules communicate with each other on an as-needed basis. Main coupling between control applications is through the common state model. While this does not rule out a hierarchical organization of control by a user, we feel that it it may suggest a heterarchical model of control (Figure 4.6).

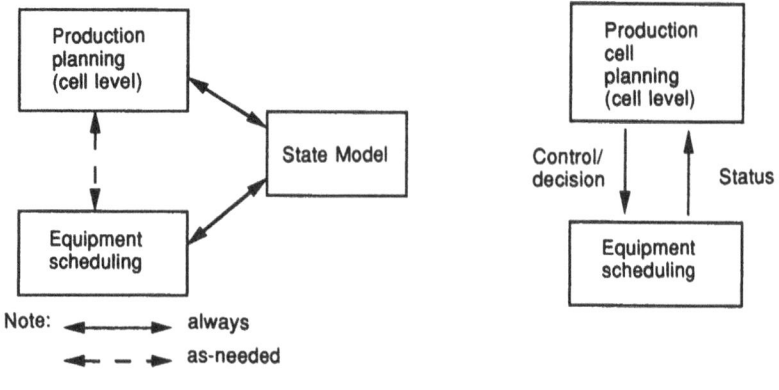

Our model: heterarchical control   **Conventional–hierarchical control**

Figure 4.6 *Two views of control.*

Modeling a manufacturing system as a collection of interacting self-contained objects is, in some ways, similar to viewing it as a distributed information processing system. Since a fundamental notion in the OO world view is to tell an object 'what to do' and not 'how to do' a task, the existence of controllers such as embedded machine controllers is not a concern of the workstation level object. This allows us the option of treating machine controllers either as a separate unit (in which case, the message intended to invoke control action is redirected to an external object) or as an embedded controller (in which case, the message is executed internally). This flexibility in modeling is needed as some machines are tightly coupled with their controllers.

This approach has the potential to address some weaknesses in the hierarchical approach. In a hierarchical system, if control at one level fails, all the levels below it cannot operate (because of lack of commands/information). In a heterarchical scheme, failure occurs only when a critical control module fails. Such failures can be handled by providing either a default or redundant control module. In practice, we expect that a combination of hierarchical and heterarchical systems may be the best solution.

4.6 Implementation perspective

In this section, we present our views on implementing the above proposal by OO principles.

4.6.1 A layered view of implementation

We propose to organize the decision system, state, and interface objects in separate layers as shown in Figure 4.7. For example, in a production control system, a workstation in the plant will have a corresponding object in the state model, decision system layer and the (state model/decision system) interface layer. The counterpart in the decision layer will have the logic for selecting the required machine to perform the production operation when a part arrives at the workstation. The interface layer will have an icon to represent the workstation graphically. An example of this implementation is that it is possible to enhance an object representing a workstation without knowing how it is displayed (i.e. through its user interface) and still use the existing display option.

Software layering is a powerful idea used in the design of operating systems (Naecker, 1988). Generally, software modularization is aimed at isolating specific software functions in separate modules so that they can be maintained independently. Software layering extends this notion by providing whole classes of software functions (e.g. database access) in a group of modules serving to insulate programs from changes in underlying support systems.

Traditionally, we have developed systems that talked to the operating system as a layer. In a way, a layer possesses the OO characteristic of encapsulation.

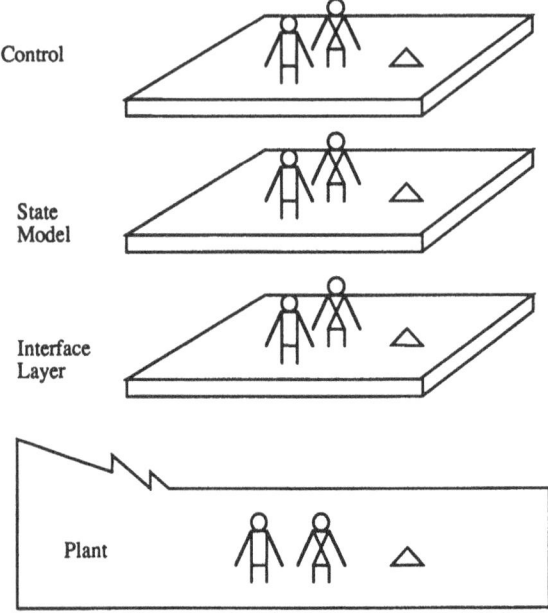

Figure 4.7   *A layered view of the general framework.*

thus enabling us to consider layered systems as loosely coupled in a vertical manner. Our framework can be implemented in three layers. Building each layer as a collection of coordinating objects minimizes any within-layer interdependencies.

It is possible to have objects at different levels of abstraction within the same layer. In particular, it is very unlikely that all the control modules will require the same level of detail in the description of the system state. For example, a planning module will probably only need to view a wafer fabrication facility in terms of areas (photolithography, implant, diffusion), while an automation module will require a description in terms of individual equipment. The state model should therefore be able to operate at different levels depending on the control logic being implemented.

Figure 4.8 shows this multi-level, multi-layer nature of a CIM system. (Interface layers are omitted to keep the diagram simple.) This scheme of layering may fit well with the marketplace realities too. The design of user interfaces and decision systems, and manufacturing domain modeling tend to be specialized fields. Therefore, we can expect to see commercial software libraries that address all these areas separately emerging soon.

4.6.2 Connection between the layers

There are two different options for connecting objects in the layers. One obvious option is that of having direct binding between corresponding objects in different layers. For example, an object in the display layer whose function is to display

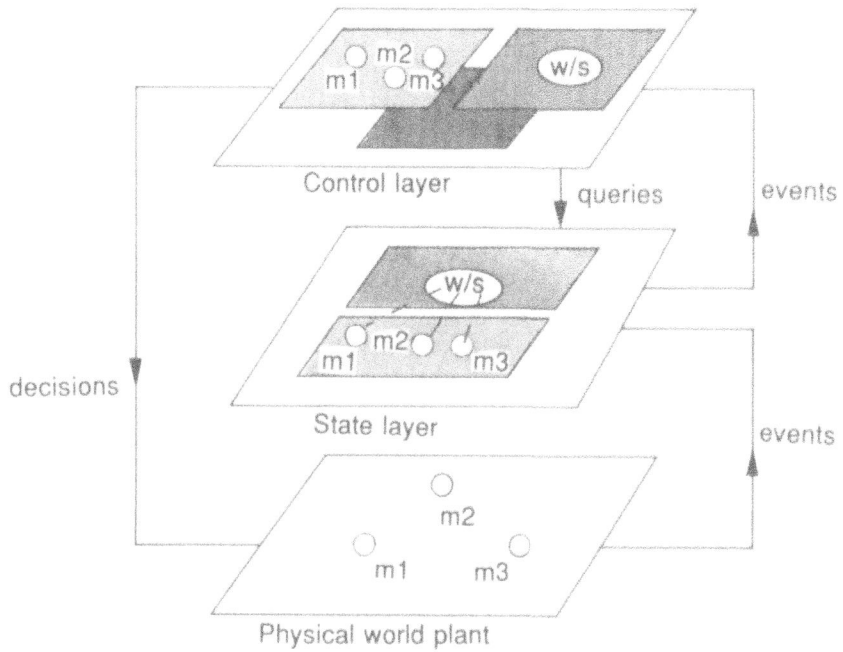

Figure 4.8 *Existence of aggregation levels within layers.*

a workstation will have the identifier of the corresponding workstation object. This will save some messaging overhead. These objects can be bound dynamically at run-time for flexibility.

An elegant and more general alternative is to route all communication through a single object, say, an object manager (or server) in each layer. This object manager provides naming and other directory services and communication with remote objects, synchronization, etc. The object managers communicate with one another through a 'software bus' that provides the dispatching function. The topic of software buses is discussed in detail in section 4.7.2.

4.6.4 Choice of a computational model for inter-module communications

The proposed architecture can be implemented using either a client–server or a peer–to–peer model of computation. In a client–server model of computation, processes are identified as either clients or servers. When a process is dedicated to managing a resource or providing a specific set of services, it is designated as a server. Other processes accessing this resource or services are designated as clients. Clients access servers through a prespecified, well-defined interface.

Communication between application modules (such as process planning and scheduling, or lot tracking and material handling) depends entirely on the dynamic needs of the respective applications. This often means that individual nodes or computers may function as clients (requesters of service) or servers (providers of service), or both. Such a need calls for a peer–to–peer model of computing instead of the popular client–server model.

The peer–to–peer model allows either partner in a conversation to initiate communication. This is a superset of functionalities allowed in a client–server model. The latter precludes a server from initiating communication with a client (needed in our case, for example, for the state model to inform its clients of change in status), while it is included in the peer–to–peer model (Tait, 1991). One way of implementing this model is through a queued message approach. This enables programs to direct or receive messages from memory queues. The latter may reside in a local system or a target remote system. This idea is good for handling asynchronous behavior as, once the program has sent a message, it is free to do other things.

The most popular implementation technique for the client–server model is the use of remote procedure calls (RPC). RPC is generally designed to work on a single application spread over distributed systems (Tait, 1991). RPCs are similar to subroutine calls and are relatively simple for programmers to use. They are not made as frequently as messages are used in a peer–to–peer network. In a peer–to–peer model, each node will have a copy of the software that provides the 'service' capability, instead of a central server copy residing on one node and manipulating clients on the remainder of the network.

4.6.5 Application design by assembly

An ambitious dream of some of those involved in OO research and development is that of being able to construct all applications as an assembly of pre-fabricated objects. One may call this a software 'design by assembly' approach. It is an interpretation of a basic principle of OOP; that is, the division of large monolithic applications into smaller units of modularity, i.e. classes. But we cannot expect that all software for a typical CIM project can be constructed from low-level software objects such as arrays, sets etc. These objects can be used to build an OO model of the application. This OO model then has to be supplemented with application data and other utility objects such as user interface objects to complete a useful application.

A CIM system will have to be built from many functional units, which themselves may contain many more lower level reusable components. Najmi and Lozinksi (1989) discuss a shop-floor scheduling system that was constructed by enhancing objects from the library, BLOCS/M (developed at U. C. Berkeley). BLOCS/M (Berkeley Library of Objects for Control and Simulation for Manufacturing) is a set of objects designed to model manufacturing systems. It was constructed from a generic software object library named ICPak–101 (supplied by Stepstone Corporation with the language Objective-C). These assembly relationships are shown in Figure 4.9.

We can view CIM applications that are to be built, as a set of models (state, decision logic and interface). Each model represents a specific set of inter-

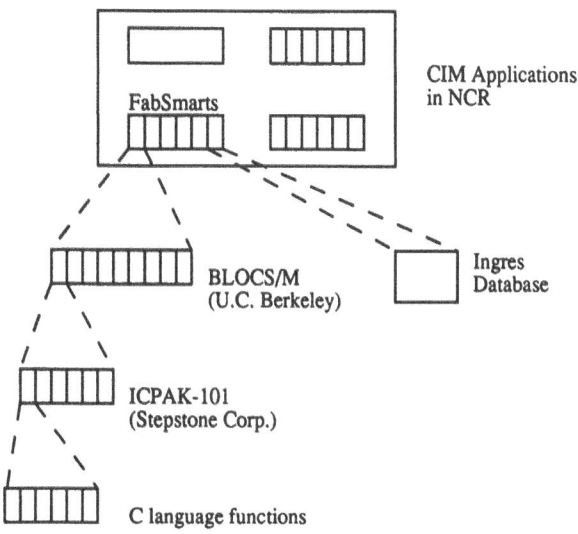

Figure 4.9 *Example of an application consisting of levels of software libraries.*

actions among a collection of object classes. At the application level, one gets the benefit of all the analysis, design and testing work done on all the models (as well as object classes). Therefore, the system designers can just concentrate on the analysis, design and testing that is specific to their applications.

4.6.6 Synergism in applications

Sometimes, when different techniques are applied to old problems, we get surprising side-effects that make quite significant contributions. Two examples are discussed next.

Control and simulation systems

An additional benefit of implementing the above proposal as an object-oriented system is in the synergism to be derived between control and simulation systems. The issue of control of a physical system is closely related to discrete event simulation, as the latter is a popular tool used to validate and/or implement the former. Just like control systems, simulation requires a model of the real world. Ideally, one would like to be able to share for simulation purposes the state model developed for control purposes (or vice-versa). The applications may share both the structure and the data. One such example is an application where one would like to simulate the future activity of the factory, starting from its current state. Also, since simulation is used to validate control strategies, control modules implementing these strategies have to be developed. Again, sharing these modules between the simulation and the control tool should be a goal. These goals can be met if we view the simulation as an event generator, which parallels the behavior of the physical system.

In this approach, the physical system and the simulation module are interchangeable, as viewed by the state model and the control modules. We can visualize the following:

1. When monitoring a real world facility, the state model is updated by events generated by the physical system, and control module implements their decisions in the physical world.
2. In simulation mode, the events sent to the state model are generated by the simulation module. Control module implements their decision in that module. For example, the control module will send the message 'Create new production task' to the simulator. In turn, the latter will send the message 'Production task started' to the state model, which does not need to know whether that message came from the real world or a simulated one.

User interface

Dealing with common objects viewed with the same philosophy may lead to the development of a commonly conceived manufacturing interface related to the way people perceive machines, parts etc.; that is, based on a common set of objects. This has the potential to have every manufacturer's software run the same way, thereby cutting down learning time for the users. This is not an unrealistic picture, as we have seen similar things happen in the past. We do not have to take fresh driving lessons when switching automobiles from, say, a Toyota Corolla to a Ford Escort, or vice-versa.

4.7 Working with existing applications and future extensions

It would be quite unrealistic and unreasonable to expect factories to reimplement all applications in an OO manner. We need to be able to provide for easy migration and also co-existence of new OO systems with existing applications that do not follow OO thinking. We also need to provide the means for conversion and growth in new applications.

4.7.1 Integration, conversions and future extensions

One may encapsulate existing applications that do not conform to an OO model. OO applications can talk to others by exchanging messages. The implementation technique necessary for this involves building object wrappers. This enables the integration of OO applications with other existing applications. This is an important issue, as pointed out by Lozier and by Wilczynski and Wallace in Chapters 9 and 10 of this book.

There is no guarantee that the open systems of tomorrow will follow the specifics being discussed today. We also believe that our strategy of achieving communication between modules via messaging (without implication of any physical or logical location of the application modules) provides considerable flexibility in accommodating future changes.

The basic mechanism needed to ensure the integrated operation of heterogeneous applications seems to be that of a 'software bus', similar in concept to the bus used to connect board-level modules in computers. This software bus should allow free communication among the modules without posing many constraints on them. The software bus is presented in further detail below.

4.7.2 The software bus

The purpose of the software bus is to manage communications between objects that span applications, i.e. application to application, or that use some common service provided by a database, etc. It does not manage objects themselves but enables communication among them. Since this bus operates above the standard communication protocols, we do not see any potential conflicts arising from protocol standardization.

The analogy of the bus relates to its hardware counterpart only in the sense that it acts as a super signal matrix to provide interconnection between objects. In order to be practical, it requires considerable support services to manage communications. One such service is a directory service that knows the class hierarchy and locations of objects used in each application and shared resource. It includes a unified name service to provide interoperability. Other services include dispatching of message requests, synchronization, exception handling, security mechanisms, activation/deactivation of remote objects etc.

Though it is conceptually simple, the software bus will require significant effort to implement. There are many ways of reducing this complexity for prototyping purposes. One such method is to register only the objects needing remote access. In its simplest form, the software bus may be implemented as a software dispatcher. The work on this software bus may become redundant if the Object Management Group (OMG) (Soley, 1990) agrees on its standard recommendation for an Object Request Broker (ORB). ORB may provide most of the functionality needed for a software bus. ORB is discussed in the next section.

4.8 Related work of interest

While there has been work in the past on CIM systems architectures (Purdue University, 1989; Beeckman, 1989, to name but two), we have not come across any architectures designed with an object-oriented world view. But there are two significant general purpose developments: the Model–View–Controller (MVC) approach, and a model by OMG. The former has influenced much seminal work in this field; the latter is a recent development that may have great impact on future applications. The MVC approach has influenced our thinking to some extent, whereas we only came across the OMG model after writing the first draft of this chapter.

4.8.1 The Model–View–Controller (MVC) approach

We do acknowledge the contribution of the MVC approach in implementing many user interface architectures. The MVC approach was made popular by the language, Smalltalk-80. An article by Krasner and Pope (1988) summarizes this approach as involving a decomposition of applications into three sets of objects:

Model: objects used to represent the application;
View: objects used to present information contained in the model to the user;
Controller: objects that manage the interaction between the user and the application.

According to Goldberg (1990), the design of the MVC roles in Smalltalk-80 came about because of a two-stage factoring. First, the design and implementation of the domain-specific aspects are handled by the model. This is separated from the second stage, work on the user interface. The latter is divided into presentation and interaction aspects.

Our approach differs from the above in many ways. We separate the model objects into two groups; the ones that capture the system state and the ones that use this information to make decisions or take control actions. We also make distinctions between two types of interfaces. While MVC was aimed at helping the design of user interfaces in general, we aim to build applications much larger in scope.

4.8.2 OMG's reference model

OMG is an international trade association incorporated as a non-profit organiz-ation in the USA (Soley, 1990). Unlike the Open Software Foundation, OMG does not plan to build and market products that implement the standards it blesses; it will merely define specifications for members and others to follow (Dyson, 1991). The OMG's reference model is named Object Management Archi-tecture (OMA).

The key components of this model Figure 4.10 are:

1. Object services: A collection of services with object interfaces to provide basic functions for managing class, instance (i.e. creation), state, security, version, etc.
2. Common facilities: A collection of objects that provide general purpose func-tionality useful in many different applications.

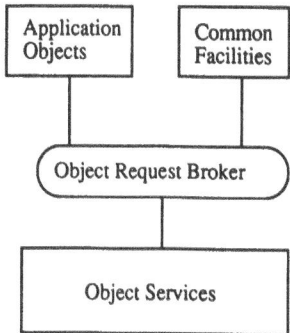

Figure 4.10 *OMG's Object Management Architecture.*

3. Application objects: These are specific to user's applications.
4. Object request broker (ORB): This interfaces applications and common facilities with required services. It has to manage the communications among objects – managing the naming, location, delivery, and synchronization services, among other things.

Interested readers are referred to Soley (1990) for the original presentation of the object management architecture, and Dyson (1991) for a concise discussion of the model.

Following the preliminary discussion on OMG's ORB, it appears that a client–server model is targeted as the implementation technique. As mentioned earlier, we are more inclined towards a peer–to–peer model of communication.

The model is at a very high level and there has been no effort by interested people to interpret its impact on manufacturing applications so far.

4.8.3 Our architecture explained after the OMG model

While the individual applications may have the structure shown in Figure 4.1, at the level of the firm, our architecture may be perceived as shown in Figure 4.11. This reference model consists of:

1. Decision or control objects:
   (a) implementing decision logic for different applications;
   (b) with one set of objects per application.
2. Interface objects: Objects implementing interfaces between decision modules and the state model, and between the state model and the plant/physical world.
3. Objects representing the state model: An object library used to model the factory; also captures the state of the factory.
4. Database objects: For storage/management of historical data.
5. Common facilities:
   (a) commonly used objects such as device drivers, e-mail facilities, graphing facilities etc.;

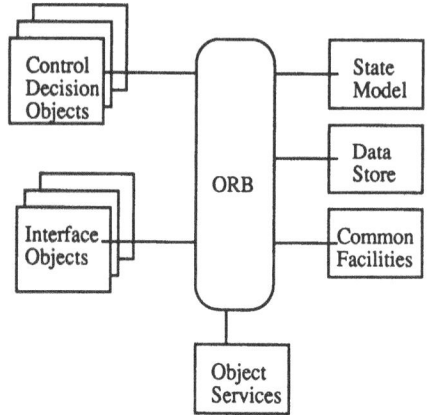

Figure 4.11   *Our framework after OMA.*

(b) interface objects which could migrate into common facilities as they become popular in usage (and standardized).
6. Object services: as in the OMA.
7. Object request broker: as in the OMA.

4.9 Limitations

The basic framework of our architecture does not reflect the organization chart of any manufacturing firm (most conventional architectures do). We do not believe that this is a serious limitation.

Since this proposal has not been implemented, there are a few questions to be answered. One of them is about its generality. Wirfs-Brock and Johnson (1990) rightly say that lack of generality generally shows up when an architecture is used to build applications – therefore, its weaknesses cannot be found until after it is designed and reused.

A question needing to be answered concerns the issue of performance. While performance may not be a serious concern in off-line modules, such as production planning, it is an issue to be examined in detail for applications demanding quicker reaction time. We hope to answer these questions after implementing this architecture in the more demanding applications of shop-floor control.

4.10 Conclusion

In this chapter we have presented a technically promising architecture as a framework for developing manufacturing software systems under the OO paradigm. This proposal for an architecture is the result of our examination of manufacturing software applications and our experience with object-oriented modeling for discrete event simulation (Glassey and Adiga, 1989), production scheduling (Adiga and Lin, 1990; Cogez, 1991), and robot control (Adiga and Leonard, 1990).

By promoting the idea of Software ICs, Brad Cox laid the foundation for work on object modularity and reusability at the object class level. This architecture is meant to promote a higher level modularity and reusability, to enable the interoperability and extensibility of applications and its major components. Each application may be viewed as a board (continuing Brad Cox's hardware analogy) that plugs into the software bus to make up an integrated CIM system.

We have defined the basic customizable components of typical CIM applications. We have also identified communication among these components (modules) in a way that parallels the real-life communication process (i.e. through events, commands, queries, etc.). This framework is to be implemented as a loosely coupled, layered architecture for CIM, one that can be implemented in an evolutionary manner. Designing to a common object model provides the logical basis for integration.

We believe that the proposed model has the makings of an open architecture, i.e. it can be easily modified (through subclassing or replacement of decision modules) or extended (by adding new applications that communicate through the software bus). All communications are through messaging. This poses fewer problems in achieving interoperability (i.e. the ability to mix applications on different vendor platforms). The encapsulation of objects has been further enhanced by a layered organization of the components of the applications. Object implementations within each layer ensure customization or programmability of applications.

We hope that interested readers will further experiment with, and enhance, the proposal described in this chapter.

References

Adiga, S. and Leonard, E. (1990) Object-oriented technology for controlling/monitoring a robot. Technical Report, ESRC 90–28, Engineering Systems Research Center, U.C. Berkeley.

Adiga, S. and Lin, W-T. (1990) An architecture for knowledge-based production scheduling systems. Technical Report, ESRC 90–1, Engineering Systems Research Center, U.C. Berkeley.

Beeckman, D. (1989) CIM–OSA: Computer integrated manufacturing–Open System Architecture. *International Journal of Computer Integrated Manufacturing*, 2(2), 94–105.

Brooks, F. P. (1987) No silver bullet: Essence and accidents of software engineering. *IEEE Computer*, 20(4), 10–18.

Cogez, P. (1991) Managing uncertainty in release decisions for semiconductor fabrication: A fuzzy logic-based approach. PhD Thesis, Industrial Eng. and Operations Research Dept., U.C. Berkeley.

Cox, B. J. (1990) Planning the software revolution: The impact of object-oriented technologies. *IEEE Software*, November, 25–33.

Dyson, E. (1991) Domain of objects: The object request broker. *Hotline on Object-Oriented Technology*, 2(8), 1–9.

Ehrlich, E., Flexner, S. B., Carruth, G. and Hawkins, J. M. (1980) *Oxford American Dictionary*. Avon Books, NY.

Glassey, R. and Adiga, S. (1989) Design of a software Object library for simulation of semiconductor manufacturing systems. *Journal of Object-Oriented Programming*, 2(4), 39–43.

Goldberg, A. (1990) Information models, views, and controllers. *Dr. Dobb's Journal*, July.

Jones, A. T. and McLean, C. R. (1986) A proposed hierarchical control model for automated manufacturing systems. *Journal of Manufacturing Systems*, 5(1), 15–25.

Krasner, G. and Pope, S. (1988) A cookbook for using the Model-View-Controller user interface paradigm in Smalltalk-80. *Journal of Object-Oriented Programming*, 1(3).

Lent, B. (1991) *Dataflow architecture for machine control*. Research Studies Press Ltd., UK.

Malakooti, B. (1989) A hierarchical, multi-objective approach to the analysis, design, and selection of computer-integrated manufacturing systems. *Robotics and Computer Integrated Manufacturing*, 6(1), 83–97.

McLean, C. R. (1987) Interface concepts for plug-compatible production management systems. *Computers in Industry*, 9, 307–318.

Naecker, P. (1988) Software layering on VMS. *DEC Professional*, September, 38–43.

Najmi, A. and Lazinski, C. (1989) Managing factory productivity using object-oriented simulation for setting shift production targets in VLSI manufacturing. Proceedings of the Autofact Conference, Society of Manufacturing Engineers, November 3–1: 3–14, Detroit, MI.

Purdue University, CIM Reference Model Committee (1989) A reference model for computer-integrated manufacturing from the viewpoint of industrial automation. *International Journal of Computer Integrated Manufacturing*, 2(2), 114–127.

Palm III, W. J. (1986) *Control systems engineering*. John Wiley & Sons.

Soley, R. M. (ed.) (1990) *Object management architecture guide*. OMG TC Document 90.9.1, OMG Inc., Framington, MA.

Tait, P. (1991) Message passing holds the key to distributed computing. *Technology Review*, Spring, 23–27.

Wirfs-Brock, R. J. and Johnson, R. E. (1990) Surveying current research in object-oriented design. *Communications of the ACM*, 33(9), 105.

5 Prototyping object systems and reusable object libraries

S. ADIGA and M. GADRE

5.1 Introduction

Since object-oriented software systems consist of a collection of interacting objects, a key activity in the design of the system is the one of defining it in terms of those objects. As Power (1987) reports, it is an accepted fact that finding the right objects for the design of a program is not an easy task. In addition, the act of mapping specifications to objects usually does not work the first time around. Some even believe that a class needs to be implemented at least five times before it is acceptable (Thomas, 1989). Designing objects for reuse is a harder task, as these objects may be used in more than one application.

Many authors have proposed different methods for designing software objects but none has found widespread use. Many rules of thumb and notational aids are available (Winblad *et al.*, 1990; Ormsby, 1991). In this chapter, we review some existing methods for designing object-oriented software and then discuss an approach that we have developed during our design work. We also discuss the basis for our approach and describe it in detail through an example.

5.2 Review of existing design methods

In a short period of time a number of approaches for the object-oriented decomposition of systems have emerged. In this section we will provide a brief review of the salient features of these approaches. The reader may note that all these methods are, generally, hybrids of existing software design approaches sprinkled liberally with heuristics. Object-oriented design methods are still in their early stages. No standard or *de facto* standard methods have yet emerged. We now briefly review the salient features of a few representative approaches to the design of object-oriented systems.

5.2.1 Object-oriented design (OOD)

Grady Booch (1986) was one of the first to present a method for designing object-oriented systems. The method is now offered as a CASE tool, called

Rational Rose, marketed by a company called Rational in Santa Clara, California. The approach was initially geared towards the Ada programming language and its idiosyncrasies. Booch (1991) suggests a design method comprising the following four steps (applied recursively):

1. Identify classes and objects. Booch provides a simple strategy using noun and verb identification from a textual description of the problem. If a number of the objects identified in this manner share characteristics, then they are candidates for being collected into a class.
2. Identify the semantics of classes and objects. In this step the protocol of each object is decided. One of the techniques suggested is to write a script for each object that defines its life cycle and behavior.
3. Identify relationships among the classes and objects. This step is viewed as an extension of the previous one, and involves establishing how different objects interact with each other within the system. Visibility decisions are made here and Booch considers this step as being crucial to producing good designs.
4. Implementing classes and objects. In this step the classes and their behaviors identified in the previous steps are implemented.

While starting by identifying nouns and verbs is intuitive, it will prove to be insufficient when the design requires classes that are abstract and do not immediately correspond to real-world objects. An example is the concept of 'Task' that is introduced later in this chapter. This class represents the interaction of several physical entities on the shop floor and is not obvious from the textual description of the problem.

Booch recommends the use of data flow diagrams (DFDs) for capturing the model of the system. While DFDs are a useful concept for general systems analysis, they are not quite adequate for object-oriented systems (OOS). In an OOS, data are contained within objects and do not flow in the same sense as that described by a DFD (Henderson-Sellers and Edwards, 1990). Communication in an OOS is via messages (which may carry data, transfer control, request services or simply inform the receiver of the occurrence of an interesting event).

While Booch considers Step 3 as being crucial, he does not provide any organized techniques for obtaining an acceptable output and recommends the use of CRC cards as proposed by Beck and Cunningham (1989).

5.2.2 Hierarchical object-oriented design (HOOD)

Hierarchical object-oriented design (HOOD) is a derivative of Booch's OOD approach and was developed under contract to the European Space Agency (Ormsby, 1991). This method also suffers from vagueness, especially on points that it considers 'crucial'. While it points out that moving from requirements to design is critical, it has no guidelines on requirements analysis itself. HOOD executes a number of basic design steps in which each object is broken into its constituents. The techniques recommended are the same as Booch's, i.e. noun identification from textual descriptions of the problem. It also relies on an Ada-like process definition language (PDL) for further formalization of the objects (Winblad *et al.*, 1990).

Since HOOD is developed primarily for use with Ada, it has no concept of classes or inheritance. This limits its use in today's environments. Other drawbacks are discussed in Ormsby (1991).

5.2.3 Layered object-oriented analysis (OOA)

Coad and Yourdon (1990) propose an object-oriented analysis (OOA) scheme that relies on layering the model. First, the method provides heuristics for identifying objects from the specifications. Then the method seeks to identify *structure*. This step consists of applying the classification and aggregation constructs used in data modeling to the objects identified (Potter and Trueblood, (1988) give more information on these constructs). This also helps to identify class hierarchies. Next, the method identifies attributes and methods.

As pointed out in Winblad *et al.* (1990), this method is simple and intuitively appealing but hardly helps the designer with messaging or polymorphism.

5.2.4 Object-oriented structured design (OOSD)

Object-oriented structured design (OOSD) was proposed by Wasserman *et al.* (1989). It is an attempt to tighten up the notation that is part of object-oriented design. It is seen as an 'architectural' design method (Ormsby, 1991). This method is based more on structured design than on object-orientation. OOSD relies on the user to identify classes and methods from initial specifications (Winblad *et al.*, 1990). Its emphasis is more on providing a complete notation rather than on proposing any one method. As pointed out by Ormsby (1991), the reason for this could be the desire to tune the method to a CASE product called Software Through Pictures marketed by IDE, for which the authors work.

5.2.5 Information structure diagram

This method was proposed by Shlaer and Mellor (1988). It consists of the following phases:

1. analysis phase;
2. external specification phase;
3. system design phase; and
4. implementation phase.

The method is based on a combination of entity–relationship (ER) analysis, dataflow diagrams and state transition analysis. Since the authors explicitly recommend the use of ER for modeling the system data, this method is the only one that has a coherent approach to identifying objects and classes.

While this method is more comprehensive than some of the others in terms of recommending specific guidelines and notations, it still falls short in terms of capturing some constructs specific to object-orientation. For example, as mentioned earlier, data flow diagrams are limited when it comes to identifying methods. In object systems data does not flow, messages are exchanged. As a result, this method is not convenient for representing messaging, capturing polymorphism or the essence of object systems – that of encapsulating data and procedures.

While state transition diagrams are appealing for simple systems and real-world physical objects, they are unwieldy and unpredictable when the number of states is large or when the object is not guaranteed to have a finite number of states. The burden is on the designer to identify *up-front* what the states of an as yet unimplemented system (or incompletely designed and specified object) will be. And since requirements usually define the expected *behavior*, states cannot be entirely deduced from the requirements either.

5.2.6 Object modeling technique (OMT)

This method has been proposed by Rumbaugh *et al.* (1991). OMT consists of the following phases:

1. Analysis phase: This begins with a problem statement and an overview of the proposed system. The output of this phase is an object model that captures:
   (a) the objects and their relationships;
   (b) the dynamic flow of control; and
   (c) the functional transformation of data.
2. System design phase: This uses the model developed in the analysis phase to organize the system into subsystems and determine the overall architecture.
3. Object design phase: This enhances the analysis models and represents a shift towards computer concepts such as the algorithms to be used and so on. The classes are refined and made more efficient. Finally, the subsystems identified in the system design phase are packaged into modules.

The analysis model tries to capture the static structure of objects, the sequencing of interactions and data transformation. The technique used to identify candidate classes is similar to Booch's – noun and verb identification. Associations between classes and class attributes are identified next. Inheritance is determined by studying the classes. The dynamic nature of objects is captured using a combination of event and state analysis. Dataflow diagrams (DFDs) are used to identify the data transformations.

The system design phase attempts to break down a complex system using various techniques such as layering and partitioning. Finally, the object design phase attempts to move the designer towards implementation in a multi-step process. It first tries to specify operations on classes from the previous phases. Algorithms to perform these operations are designed and data access is improved. Recommendations for design optimization are provided.

In general, Rumbaugh *et al.* (1991) is readable and provides many useful ideas, especially for the decomposition of complex systems. Their method is supported by a CASE tool called OMTool that implements their method. However, their method suffers from many of the same drawbacks as the others and has some that are unique to it.

OMT emphasizes design optimization as well as tuning object structures for specific algorithms. The algorithms in turn are designed and implemented based on the requirements for the current system. Such an approach may lead to systems optimized for the task at hand but difficult to enhance. We feel that it may not achieve the full benefits of reusability.

Another problem is the use of dataflow diagrams. OMT as a result fails to help the designer identify polymorphism. The reliance on old structured analysis and design techniques forces the authors into mixing paradigms to capture object characteristics. This is particularly the case in the object design phase. In general, there is a tendency to mix bottom-up approaches prematurely with top-down ones.

In the analysis phase, OMT uses a combination of event and state transition analyses to identify dynamic behavior. There is a real potential for confusion in mixing these modes. This is especially so in the case of objects which are still in the process of refinement. As we discuss in section 7 of this chapter (State Transition Analysis), an event analysis leads to an identification of the states and vice-versa. However, events usually have the advantage of being independent of the system. States, especially object states, do not. And states for an object that is still in the process of being identified and designed are hard to pin down. Forcing the designer to mix these techniques this early in the method may lead to confusion.

5.2.7 Summary

While many methods have been proposed, we have found that most of them lack the features we would like to have to guide the development of object-oriented systems for manufacturing. These desired features include the ability to proto-type rapidly, an event orientation and an object-orientation at all levels of system design. Most methods are also quite lacking in providing techniques for identifying object-oriented constructs such as polymorphism and messaging. All the approaches rely on mixing existing techniques such as ER diagrams, state transition diagrams and data flow diagrams. However, none of these techniques was developed for use with object systems and therefore must be enhanced.

We also observe that most of the approaches (information structure diagram excluded) depend on a noun- and verb-based identification scheme as a way of getting the design process started. These methods provide long lists of heuristics to eliminate data redundancies in the objects identified. Upon closer examination this is similar to a basic ER analysis. By avoiding the explicit use of ER the methods end up requiring special notation and rules of thumb. This avoidance also leads to the need for specialized CASE tools (that implement the authors' particular set of heuristics). We advocate the use of ER quite explicitly. As a result, a mature method is immediately made available without any new notation. The other benefit is that users of our method need not spend a lot of money on expensive and proprietary CASE tools; many reasonably priced ER tools are available on personal computers.

5.3 Rationale for our approach

A fundamental concern of ours is to avoid over-structuring the problem which could result in over-management and over-specification. This in turn, as pointed out by Lawson (1990), may lead to volumes of human and/or computer generated documentation that nobody wants or bothers to read. Our approach is based on the following considerations.

1. We believe that good systems evolve, and experience with earlier versions leads to improvements in the future releases. (The spiral model of software development has been formalized by Boehm (1988).) This is particularly true for the development of reusable, domain-specific classes. This points us to a need for fast prototyping, which allows the system developers to learn, improve and evolve the system.
2. Prototyping is particularly relevant for the development of reusable libraries of objects. Generic libraries such as ICpak 101 provided by Stepstone (1989) or user interface libraries can be of limited help. Such libraries are useful for 'utility' work such as array management, graphical displays and so on. However, most projects require '*domain-specific*' libraries that can be used to accurately model the application domain. Currently, domain-specific objects are not available 'off the shelf'; they must be developed in-house. In order for the library to be robust and truly reusable, implementation experience and feedback is required. Rapid prototyping provides this experience.
3. We also believe that computer applications are fundamentally only simulations of the real world. Simulation can be a useful paradigm for the development of object-oriented systems. Further, this approach provides a 'modeling' perspective to the design process.
4. A manufacturing software system can be viewed as a set of functional modules or process 'objects' communicating with each other through the exchange of inter-process messages. The aggregation of the functionality represented in the modules, as well as their system of communication, defines the system behavior.
5. While the above considerations lead to an identification of the methods to use in order to model object behavior under different situations, the identification of these situations that trigger system behavior is necessary. We believe that

an event-based analysis is the most convenient to comprehend and apply. Focusing on events makes it easier to relate software functionality to real-world events. It also ties in with our view of computer applications as simulations of the real world.

6. Since objects encapsulate both data and behavior, it is essential that an object-oriented design method be able to capture both aspects. The entity–relationship (ER) approach can be used to model the structural associations between objects, while *message flow diagrams* (explained below) lead to the dynamic relationships.

5.3.1 A conceptual view of our approach

In our approach, we treat the entire process of analysis and design as development and refinement of a series of three models. Each model is built based on knowledge of the one built previously. Together, the three models describe what a system does, with minimal constraints on how it should be implemented. These three models are as follows.

1. *Process model*: This represents the main processes to be examined or independent processes comprising the system to be built. These processes may be viewed as the basic subsystems.
2. *Data model*: This helps us to identify the basic entities (and entity sets) that comprise each subsystem and their properties (attributes). The data model is used to identify necessary objects (and object classes). The attributes help us to identify instance variables needed in objects.
3. *Event model*: Events cause changes to the state of the system. These changes are identified in terms of behavior needed in objects to respond to the demands of the event considered. The event model captures the message interactions between objects for all important events. It serves to identify computations or methods needed in objects. If the instance variables identified earlier are found inadequate to support the data requirement of the methods, new instance variables are added to the respective object(s).

Binding between the models is provided by an analysis logic based on a simulation of discrete events in the real life processses. Thus, discrete event simulation may be considered as a design metaphor in our approach. A common set of objects required for building all the processes in the system can be obtained by a union of all the objects in individual process models.

5.3.2 System partitioning through process modeling

Using the functional decomposition approach we can break the requirements into a set of processes. This will help to localize functionality as well as identify the important data. We prefer the functional decomposition approach at this stage for the following reasons.

In our view of object-oriented systems, a process can be viewed as a system level 'object'. Just as a software object encapsulates both data and functionality, so does a system level process object. In this manner, the entire system can be viewed as a set of objects that interact via (inter-process) messages and together provide the desired system level functionality.

It is usually the case that the new system under consideration needs to integrate existing functional modules into it. The manner in which this can be achieved in the context of the design process is through the use of functional decomposition. A functional decomposition helps the designer move easily from the system requirements – which are usually in the form of what the system is expected to do – to a set of functionally independent modules that communicate with each other. The communication is captured in the message flow diagrams.

5.3.3 Data modeling with entity–relationship diagrams

The functional analysis will yield a set of independent functional modules that constitute the system. However, the data that represents the underlying model needs to be seamlessly integrated into the object model.

The ER approach (Chen, 1976), used in data modeling for databases, provides a convenient formalism for identifying both physical and conceptual entities. It is commonly used to represent the static or structural relationships among entities in the real world. Its use is quite widespread among database designers and students of data management. It can be used not only as a modeling tool but also as a design tool for a software specification independent of any implementation language. Since the concept of an entity is focused primarily on data management, we will have to define certain abstraction rules to relate the entities and relationships to classes of objects in the object-oriented paradigm.

An entity–relationship diagram (ERD) will identify the structural relationships between the interacting objects. Since objects encapsulate both data and functionality, we need a mechanism to identify the dynamic relationships and interactions. This is accomplished by event modeling and analysis.

5.3.4 Event modeling with message flow diagrams

Event analysis

We take Kowal's (1988) view that 'a system is built to respond to events'. It is our belief that an event based analysis is perhaps the most intuitive for the designer and developer as well as the user of a manufacturing system. The user can relate easily to a view that sees the system as responding to a set of events generated by his or her actions rather than, for example, viewing the system as going from State-A to State-B (as would be the case in a state transition analysis).

In addition, event analysis ties in with our view of computer applications being simulations of reality. In a discrete event model (see, for example, Law and

Kelton (1982)), the events that trigger changes in system state need to be identified. As described by Shaw (1987), real world modeling consists of objects and
events. During an event, objects interact and as a result some may change their
respective states. Event analysis leads to an identification of these occurences.

The following definitions are taken from Kowal (1988): 'The purpose of a
system is to provide the necessary response whenever an input stimulus, a time
event, or an anomalous condition is recognized by the system.' He defines an
event to be an 'independent occurrence outside the system, an occurrence of time,
or (in real-time systems) an occurrence detected inside the system that causes
activity in, or results from, the system.'

Most of the important events can be identified from the system requirements.
The requirements usually state the expected user interactions with the system
(note that user may mean both a human user and/or another process or program).
This cannot be said for object states.

An event analysis coupled with message flow diagrams will lead to an identification of the states of the system and objects. State transition diagrams may be
drawn at this stage if so desired. An example illustrating this is shown towards
the end of this chapter.

Message flow diagrams

Communication in an OOS is via messages. This inter-object communication
involves combinations of data transfer, control flow and service requests. The
sequence in which these messages are sent and the objects that are involved is
represented in a *message flow diagram (MFD)* – see also Glassey and Adiga,
(1989) and Adiga and Gadre (1990). An MFD is simply a network where the
nodes represent objects and the arcs connecting them represent messages. The
arcs are numbered in the same sequence in which the messages are sent.

It is important to note that a message flow diagram (MFD) and a data flow
diagram (DFD) *are different*. Data flows are pipelines along which data and
information of known composition passes between components of the system. A
DFD documents this flow of data (Kowal, 1988). However, in an OOS, data are
contained within objects and *do not flow in the same sense* as that described by
a DFD (Henderson Sellers and Edwards, 1990). An MFD is essential for captur
ing the communication within an object system.

Since our expanded view of objects includes processes at the system level, an
MFD drawn at this high level will identity both the transfer of information as
well as enhance the functional requirements on the receiving process. A message
received by a process may be one of the events triggering changes in the objects
that make up that process. We believe that DFDs are not sufficient to capture
these features needed in an object-oriented system.

At the object (class) level, the arcs (messages) map into methods that are to be
provided on the interface of the receiver object. Based on the analysis of the
event and system behavior, the functionality of and return values from the method

can be determined. This may lead to a revision of the list of instance variables of that object.

MFDs can get complex. A message to an object can result in a cascade of messages from that object to others. A useful guideline in developing MFDs is to have one MFD per single event. This helps to define the context for an MFD as well as keep message propagation at the same level of abstraction.

In our MFDs we use the syntax of Objective-C (Stepstone, 1989) for the messages. This is only for the sake of brevity, convenience and because we like Objective-C syntax for its readability. Users can use any nomenclature of their choice. The message syntax follows. If unspecified, the default variable type is **id** (see Appendix).

[**receiverObject message**: (arguments)];

Multiple arguments can be passed using the ':' symbol. For example:

[**window moveToX**: (int) **andY**: (int)];

5.3.5 Class hierarchies

Identifying a hierarchy is necessary as it is an important aid in determining the reusability of the resulting software object library. The ER models do not specify hierarchical relationships explicitly. Heuristic rules as well as experience with prototypes are required to identify them. However, extensions to the ER model discussed in modern textbooks on database analysis (Date, 1990; Ozkarahan, 1990) provide suggestions to identify 'generalization' hierarchies.

Class hierarchies are abstractions in the artificial domain that is constructed out of the problem domain that we are modeling. The need for abstract classes is quite significant in solving complex problems. In a production scheduling application using discrete event simulation, we had to define 'future event' objects – breakdowns, event calendar etc – a mix of real and unreal objects which one would never gain intuitively if one plans a hierarchy only from the real problem domain (Glassey and Adiga, 1990).

While the concept of inheritance lends much power to OOS, it is not easy to implement it in a manner that will lead to reusable software. We agree with Scharenberg and Dunsmore (1990) that there is a natural evolution of class hierarchies. That is, there is no need to plan for a hierarchy; it evolves when the domain is modeled in the form of ER diagrams.

5.3.6 Summary

Considering the above discussions, it is appropriate that we should look at a structural decomposition of the domain to identify objects as well as the events that the objects go through in the proposed system. We need to relate the numerous changes the system is expected to go through to the composition of objects.

The approach must be simple enough to support rapid prototyping and the many cycles of the design process.

This has led us to believe that a systematic approach based on:

1. a functional decomposition and isolation of the high level requirements;
2. an identification of key abstractions using a data modeling approach such as ER; and
3. an analysis of the events that the objects participate in,

may lead to a good conceptual design. It would also keep the method simple as well as easy to comprehend and apply.

While the above view is useful to explain the conceptual basis for our approach, more elaborate and discrete set of steps may be more useful for design professionals. The design process consisting of the activities to be performed is described next. The activities are grouped into three phases.

5.4 Our design approach

In this section, we present a hybrid approach that advocates analysis and design of the target system using functional decomposition, data modeling techniques and relevant event analyses to develop specifications for software objects. There are three phases in the development of a prototype of an OOS. We now describe these phases and the important steps to be followed within each of them.

5.4.1 Description

Phase 1: System partitioning into subsystems

We propose a combination of functional decomposition coupled with relevant event analysis to break the system requirements down into a set of functional modules or '*processes*'. This step also identifies the high-level inter-process communication that needs to take place between these modules.

Any system has a set of functions that it needs to perform. Using functional analysis we can decompose the system into a set of independent functional modules. Usually, the new system to be developed needs to integrate the functionality that already exists in software (developed in-house or third party) that the enterprise is using. For example, if the manufacturing facility is already using a sophisticated statistical quality control (SQC) package that has been bought from an independent vendor, it is wiser to integrate the SQC package into the new system rather than develop it from scratch. Functional analysis makes it possible to integrate existing software modules into the system that we wish to develop.

In manufacturing systems, it is quite likely that we will need asynchronous processes (modules) that operate independently of each other but communicate with each other via inter-process messages. For example, the automatic data collection that takes place on the shop floor should not be affected by the fact that someone is using the Costing module at the same time. However, if the SQC module determines that certain additional data is required, it should be able to communicate this request to the data collection module.

Based on the system requirements, we can determine the important events to which the system must respond. An event may require functionality that is distributed between processes and so may trigger communication between multiple processes. The high-level communication between these processes is captured in message flow diagrams (MFDs) that pictorially represent the inter-process communication.

According to Coad and Yourdon (1990), in functional decomposition, analysts end up with system, subsystem, function and sub-function levels. When using functional decomposition in the context of object-oriented systems, we recommend stopping after the subsystems are identified and then going on to Phase II (described below).

It is important to note that even if we are not designing a multi-process application, this phase is quite relevant. A single process application usually needs to satisfy different functional requirements. A functional decomposition coupled with event analyses will help to identify the different sections of the process and the communication (in the form of data and/or command flows) among them.

The output from this stage will be a set of functional modules, a set of important events that the system must respond to and a set of MFDs outlining the high level communication between these processes.

Phase II: Object library identification

Once we have identified the processes or main functional subsystems, it is necessary to abstract the requirements that are common among them. Since we wish to promote software reuse, it is essential that the software that is developed in order to meet these common needs be in the form of reusable object libraries. For example, a common need for different processes running on the manufacturing floor is a model of the manufacturing facility. This model will be essential for data management, costing, scheduling and other functions. A common user interface library is another necessity.

The output of this stage will be specifications for a set of object libraries that we need to develop in order to satisfy the requirements of the system. The specifications must define the problems and objectives of the library and applications to be developed with it. The scope of each library should also be clearly spelled out. This will determine the level of detail that must be obtained with the software objects that make up the libraries.

This stage is also useful in identifying software layers. The libraries can in some cases be viewed as providing support for these layers; for example, a user interface layer, an inter-process communication layer, and so on.

Phase III: Software library decomposition

In this phase of the design we propose a combination of a data modeling technique and event analysis to decompose the library specifications into reusable software objects. This step is crucial to the enterprise developing the system because it is at this stage that the domain-specific reusable components are designed and developed. The steps involved in this stage are as follows.

1. Based on the library requirements, define its objectives.
2. Study the application environment to collect information needed to model the problem domain.
3. Identify entities, relationships, and attributes. Develop a normalized entity relationship diagram (ERD).
4. Based on the above diagram, abstract objects (or classes), and instance variables.
5. Identify the main events of interest and the objects affected by these events.
6. Draw a message flow diagram (MFD) for each event to identify the behaviors exhibited by the interacting objects. This activity will suggest methods required by different objects.
7. Revise object definitions if needed, and consolidate specifications for all the object classes in the system.
8. Implement the design in an object-oriented programming language (OOPL).

Steps 2 through 7 may have to be gone over iteratively until a satisfactory model is obtained. These iterations will be partly driven by experience derived from prototypes and simulations developed in Step 8.

Documentation

To reuse software objects, it is essential to have detailed and current documentation. Insufficient documentation makes the potential user of a software object hesitant. He or she is not willing to use something whose behavior is not explicitly spelled out. Incorrect documentation as well as documentation that does not match the code casts a shadow of doubt on the code itself. A potential user may wonder whether the documentation or the code, is correct.

While few will doubt the importance of documentation, fewer still would want to spend time and energy maintaining numerous separate documents. We believe that all relevant documentation should be part of the source files themselves. If this documentation is written in a standardized manner, any text processing utility can automatically strip the documentation from the source files and create manual pages. We present a documentation template that users may use directly or with suitable modifications in the Appendix to this chapter.

Graphical documentation (for example, MFDs) cannot be incorporated into source files. These must be maintained separately in an appropriate, easily exchanged format (for example, PostScript). While this involves the maintenance of multiple documents, we believe that the MFDs and ERDs will not change as rapidly or as often as the code level documentation.

Documentation of an object could evolve with the analysis and design activities. For example, Wirfs-Brock *et al.* (1990) have found it useful to incorporate CRC cards – a form of documentation advocated by Beck and Cunningham (1989) – into their design process.

Summary

The steps presented in this section apply the object-oriented paradigm all the way from high-level system design down to the specifications of software objects. The designer may choose to apply only the relevant steps according to his needs. For example, if the requirement is only to design an object library then only Phase III need be applied.

5.4.2 Further discussion on Phase III (library decomposition)

Currently, as we have discussed above, reusable domain-specific code cannot be bought 'off the shelf'. It must be developed in-house and requires a deep understanding of the domain and its behavior. While such code is essential for the project, it may be quite difficult to develop as there will likely be few examples. As a result, the issues discussed above – such as the need for rapid prototyping and techniques for capturing object features – become even more relevant. For these reasons we believe that it is essential to further explain and clarify the steps involved in this phase of object-oriented system design and development.

Step 1: Define the problem and objectives

A clear definition of the problem(s) to be solved will help set the scope of the data model to be built and also help decide the level of abstractions needed in the model. For example, if the problem is to predict workstation utilization for a projected increase in load, the workstation can be treated as a set of identical machines. That is, machines need not have separate existence as objects.

If the designer has come to Phase III directly, then this definition must be developed. If the designer has followed Phases I and II, the library specifications developed in Phase II can be used as the context for this step.

Step 2: Understand the application environment

Studying the application environment helps one identify the terminology used, characteristics of the entities involved, and helps in conceptualization of the problem domain. Some of the important questions to ask oneself are:

1. What are the important entities in the problem domain?
2. What is the essential level of abstraction of the entities involved?
3. What information do we need?
4. What are the assumptions?

Step 3: Entity–relationship (ER) model

The entity–relationship approach to data modeling was introduced by P. P. Chen (1976). It is a conceptually simple and attractive design approach which has been widely used in the logical design of databases. A detailed discussion is beyond the scope of this chapter; readers are advised to look up any standard textbook or paper on database design (such as Hawryszkiewycz, (1984); Potter and True-blood, (1988) or Date (1990)). An introduction to the basic concepts is given here to enable the readers who are not familiar with ER to follow our method.

The ER aproach relies on three constructs: entities, attributes, and relation-ships. An entity is anything of interest to the organization (or problem at hand) about which data can be collected and stored. An entity can be a physical thing (i.e. a person or a computer), a concept, an organization, or an event of interest. An entity type (or set) is a group of entities satisfying certain criteria or possess-ing the same characteristics. For example, in a shop floor production supervisor's office a partial list of identifiable entity sets might be the following:

1. Operators – human beings;
2. Machines and wafers – real objects;
3. Maintenance schedule – a written document;
4. Start of production – an event.

An attribute is a characteristic of interest in an entity. Attributes comprise the data that is collected and stored about entities. Attributes of the entity, Operator, might be: name, address and the list of machines that he or she is qualified to operate. An entity set is represented in an ER diagram by a square or rectangular box.

A relationship is an abstraction of the association between entities. It is shown in an ER diagram by a diamond shaped box containing the description of the relationship. Relationships symbolize the connection between entities. For example, Operator *operates* a Machine. The relation 'operates' implies simulta-neous occurrences of instances of two entities, one each from Operator and Machine. A relationship type (or set) is a classification of relationships based on certain criteria or characteristics. In a generalized ER model, the relationships can also have attributes. For example, the time to process a part on a machine could be dependent on a combination of machine and operator (particularly when many operators may operate many different machines).

Step 4: Identify classes and instance variables from the ERD

We find that distinct entity sets map directly into distinct object classes. The objects (instances of the class) represent the entities. It is important to consider the entities as abstract but meaningful 'things' that exist in the user enterprise. Another point to remember is that only those relationships having attributes can be considered candidates to be mapped into object classes. Attributes are transformed directly into instance variables of the corresponding classes. Basic heuristics used in the logical design of databases (see, for example, Teorey *et al.* (1986)) are useful in the mapping of entity sets to object classes.

We note that there are basically three ways in which objects relate to each other (Potter and Trueblood, 1988; Ozkarahan, 1990):

1. through an association relationship (shown in most ER diagrams);
2. as part of an aggregation structure; or
3. as a type of specialization of a generalization structure.

The generalization–specialization relationships (for example, equipment–lathes) lead to class hierarchies. The part-of aggregations (for example, gear box, steering wheel etc. are parts of an automobile) lead to composite objects.

Step 5: Identify important events and the objects involved

Systems are built to respond to events (Kowal, 1988). An event of importance will result in the application of a stimulus to the system that the system is intended to be capable of responding to. For example, in the case of a graphical user interface application, the user clicking on the mouse button will be an event of importance to the system. The objects involved will be the window that is displayed as well as the object(s) dedicated to processing the mouse input. (See Kowal (1988) for additional details.)

Step 6: Draw MFDs for each event and identify methods

An event will trigger a change in the states of one or more objects in the subsystem. This change may trigger a cascade of messages to other objects, which may in turn trigger further changes in state and associated behavior. For each event, a message flow diagram detailing this communication is drawn. The arcs, which represent messages, translate into methods on the receiving objects. The desired object behavior in response to these messages translates into the functionality represented by the methods. At this stage we identify the set of methods and their functionality represented by the methods. At this stage we identify the set of methods and their functionality associated with each event as identified in Step 5.

Each diagram must contain only essential actions that are triggered by a given event. Communication between events is through changes in the state(s) of the system (or a set of specific objects).

Sometimes we may not be able to explain an event with the given objects. This indicates an omission in the earlier analysis and signals the need for a new object or objects.

Step 7: Revise and consolidate object specifications

Now we consolidate the information derived from the previous steps. As mentioned above, identifying methods and their functionality from the MFDs can lead to revisions of the instance variables identified in Step 3. In the process of consolidation, we may discover inconsistencies and/or redundancies in the classes, instance variables and methods identified earlier. Some important data may have been missed or the overall functionality may not correspond exactly with the requirements. As a result, we may have to iterate Steps 1 through 6 to arrive at an acceptable design.

At this stage we develop specification sheets for each class. We represent this information in a tabular form that includes the class name, the list of instance variables as well as the methods to be provided on the class interface.

Step 8: Implement the design in an OOP language (OOPL)

Given the specification sheets for each class it is now time to convert the design into code. The prototypes and simulations that are produced will influence further revisions of the design. The experience gathered at this stage will be useful in developing class hierarchies as well. Based on the level of maturity of the system, developers can decide whether to further refine the implementation or not.

Language independence

While Steps 2 through 5 and most of 6 can be carried out without reference to any specific OOPL, it is difficult to avoid such references when deriving the functional specifications. This is also true when we are trying to abstract class hierarchies. For what may be possible in one OOPL may not be so in another. For example, multiple inheritance may not be provided in all languages. Dynamic binding may be more difficult in some languages as compared to others. In general, as we start modeling the specifics of the object interactions and behaviors the model gets more detailed, and may be influenced by the implementation language.

The design approach will now be illustrated through an example.

5.5 An example manufacturing facility

To illustrate our design method, we will describe a simplified semiconductor wafer testing facility. The example is kept simple in order to highlight the modeling method. While the method can be applied to very complex situations, we want to avoid having the complexity of the example distract us from the principles behind the method.

A wafer testing facility is a very complex manufacturing system. Hundreds of thousands of circuit elements are imprinted on silicon wafers, with the processes involving many individual operations.

We can assume that the system will read from and write to a common database (i.e. this example does not include database issues). The primary motive for the development of this system is to help supervisors improve the manufacturing efficiency of the test facility. This includes improving equipment utilization, product quality as well as the throughput of the facility. The software system in the test area must have the following major functions.

The system must provide the capability to track equipment, materials and labor in the test area. It must maintain an up-to-date representation of the state(s) of all elements of interest in this area. Central to this function is an accurate physical model of the facility.

Another important capability is production management. This is responsible for both generating feasible schedules as well as providing the user with 'what-if' capability. Because of the high equipment cost, and the potential profits from the sales of finished goods, a wafer fab aims to maximize both the number of finished goods as well as the utilization of its equipment.

These problems can usually be countered by a detailed analysis of the fab and appropriate scheduling and dispatching to streamline production. The 'what-if' capability helps the user understand the behavior of the facility under different conditions and situations as well as test the effectiveness of alternate solutions. Developing solutions often involves software design in the form of simulation or a software implementation.

The physical details of the facility are as follows. There is a set of workstations, each of which can be in any one of three states – IDLE, BUSY or INSETUP. Each workstation consists of a number of identical, completely reliable machines. While the machines are physically identical, it is possible for different machines in the same workstation to be set up for different operations.

There is a queue of wafers awaiting processing (testing) in front of each workstation. The wafers follow different sequences of tests at the workstations and join the finished goods inventory once tested. The wafers are grouped into lots, with a lot of wafers staying together for the complete testing sequence. Hence, for the sake of simplicity, a lot can be considered to be a set of identical wafers. (In reality, this need not be the case.) In order to make decisions related to resource allocation and control, there is a 'coordinator' object which is informed of every state change that occurs in the fab.

An important decision is to be made when a workstation finishes processing a particular lot. This decision answers the question of 'which lot to process next' and is known as the production dispatching decision. Different rules and heuristics have been found useful under different conditions. The choice of dispatching rules can affect the performance of the fab. Central to the production management activity is a discrete event simulation of the facility.

To ensure that the testing and other operations are within specifications, there must be some control of the process. Process control ensures that the equipment and other physical processes are operating within prescribed limits. The control may be enforced and relevant data may be acquired through machine level programmable logic controllers (PLCs). Statistical quality control (SQC) is an important element of this function. This function will feed information relating to the state(s) of the equipment and materials to Tracking.

Finally, there must be an element of cost control. This includes keeping tabs on aspects such as the equipment utilization, yield of the processes as well as comparing the cost benefits of different schedules generated by the scheduling function. The system is also responsible for managing demand for finished product and passing information relating to new orders to decision support.

To keep the analysis to a reasonable length, we will restrict ourselves to the following events of interest. The list is by no means comprehensive, even by the simplified description presented above:

1. Start of production (testing) at a workstation.
2. End of production (testing) at a workstation.
3. Start performing a setup on a workstation.
4. End performing a setup on a workstation.
5. Receipt of a new order by costing.

5.5.1 Phase I: System partitioning into subsystems

Based on our description of the manufacturing facility and the requirements of the software system we can identify the following functional modules:

1. Tracking
2. Process control
3. Production management
4. Costing

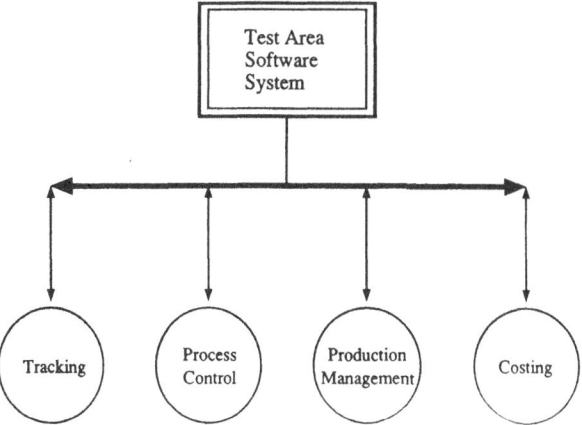

Figure 5.1 *The process model.*

Figure 5.1 shows the process model. In addition to the four subsystems identified, there is a need for a mechanism that facilitates inter-process messages. This
is the 'software bus' discussed in Chapter 4. In Figure 5.1 it is represented by
the thick double-headed arrow. Redrawing Figure 5.1 to account for this component, we obtain Figure 5.2.

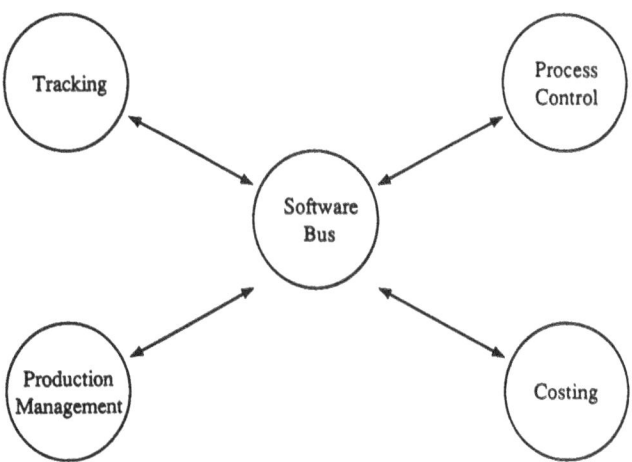

Figure 5.2   *Revised process model showing the software bus.*

We have used circles to represent subsystems (for example, in Figure 5.2) and
objects (for example, in Figures 5.11 and 5.12). This is consistent with our view
of subsystems as high-level, process 'objects'.

The five events that we will consider in this example have already been listed
above. Let us draw the MFD for one of the events: end of production at a
workstation. This is an event that is external to the software system.

It is expected that the Process Control module will be the first to know of this
event. It will be notified of this event perhaps through the PLC (programmable logic
controller) attached to that workstation. As soon as it is notified, it will send a
message to Tracking informing it of the event along with relevant information such
as the name of the workstation, the lot id, date and time of occurrence of the event.

Tracking will then store relevant data in the database. It will also notify Production Management of the event and pass it the information related to the workstation, lot and so on. Production Management must now determine what the next
action should be. This determination will be driven by various dispatching rules
coupled with the manufacturing schedule in effect for the facility. Production
Management may have to read the current state(s) of the upstream and downstream
equipment from the database in order to arrive at its conclusions. The final decision will include the lot to be processed next and the correct setup required. This
information will be transmitted to Process Control in the form of another message.

This high-level, inter-process MFD is presented in Figure 5.3. It is understood
that all interprocess messages will go via the software bus. Hence, for the sake

of simplicity, we have not explicitly shown this routing in Figure 5.3. In a similar manner, we can draw MFDs for all events of interest.

In Figure 5.3 we can see that a single message **endProduction** is used to communicate the occurrence of the event. This message will result in module-specific behavior depending on the receiving module. This characteristic of object systems, called '*polymorphism*', can be easily identified by this approach. An added benefit of polymorphism is that it can help standardize on message names and thus make the system easier to understand.

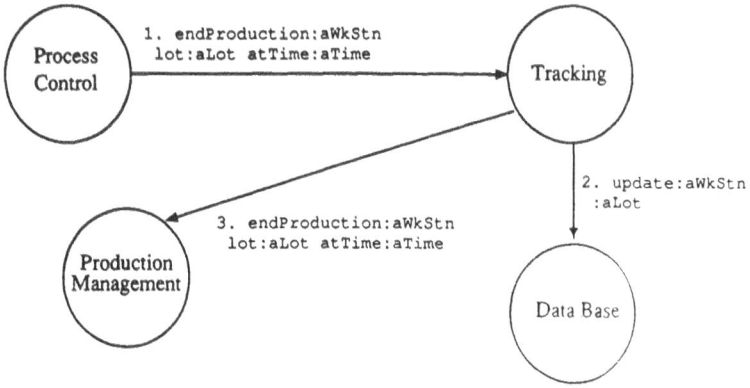

Figure 5.3 *Inter-process MFD at the end of production.*

5.5.2 Phase II: Object library identification

Based on the analysis of CIM in the previous chapter, we can see that the essential libraries to be designed relate to modeling the facility, building user interfaces and managing data storage and retrieval. Alternatively, if one examines the basic subsystems, he or she may see the need for the following requirements in addition to the problem-solving logic needed in each subsystem.

1. Process Control requires:
   (a) A model of the different activities being performed on the shop floor. This might include structural data such as routes (operation sequences).
   (b) Automation and human interfaces designed for controlling the equipment.
2. Tracking requires:
   (a) A structural data model of the manufacturing facility.
   (b) A user interface designed for setting up the structural data.
3. Production Management requires:
   (a) A model of the manufacturing facility, both structural and dynamic (i.e. what are the elements affected by a particular event, how are decisions related, what are shop floor activities and so on).
   (b) A user interface designed for the user to perform 'what-if' analyses as well as select and set up dispatching rules.

4. Costing requires:
   (a) A model of the facility.
   (b) A user interface designed for entering orders, comparing schedules and so on.
   (c) A model of the different activities being performed on the shop floor.

From the above we can see that one of the crucial requirements is a model of the facility that represents both structural and dynamic relationships. Also essential is a common user interface library that can be used to construct the different screens required by the modules.

A user interface library is something that can be purchased off-the-shelf. Depending on the target platform(s) there are many vendors that supply reusable, portable libraries. These provide facilities for creating and managing elements such as windows, scroll bars, list boxes and push buttons. Once a consistent 'look and feel' for the product has been agreed upon, this functionality can be encapsulated by the different modules to produce the desired screens.

However, an object library that can be used to model the facility is not something that can be purchased. It involves far too many domain-specific features and requirements. This library must be designed and developed in-house. As a result, we will focus mostly on this library in this example.

Specifications for an object library to model the facility

The object library to be developed must be capable of capturing the structural details of the facility as described above. In addition, it must be capable of responding to the (specified) different events that the system is subjected to. The components of the library must have a concept of 'persistence' – i.e. the ability to store themselves and recreate themselves from the database. The library should allow the users to model different dispatching rules and heuristics and use and test them with minimum effort. It should support discrete event simulation analyses of the facility for the purpose of providing what-if analyses and making dispatching and scheduling decisions. It must be remembered that the primary reason for the development of the software system (and thus the library) is to provide support for efforts to improve equipment utilization, product quality and throughput.

5.6 Object-oriented library for modeling the facility

This is Phase III of our approach. Based on the above description of the facility we can design and implement an object library to model the system.

5.6.1 Step 1: Define the problem and objectives

We are required to develop an object model of the facility. This model will be used to improve the equipment utilization as well as throughput of the test area.

Different dispatching and scheduling rules need to be tested in order to achieve these results. Collecting statistical data and reading and writing to the database is an important aspect of this activity.

5.6.2 Step 2: Understand the application environment

The important real world entities in our problem domain are workstations, machines, operations and operation sequences, lots (of wafers) and a coordinator.

If we follow the 'next event time advance' approach to discrete event simulation (Law and Kelton, 1982) we will need an object that functions as an event calendar. This is simply a stack of future events ordered by their event activation times. When the simulation clock is advanced to the next activation time, the corresponding object is informed. We name this class **Timer**. The **Timer** increments the simulation clock to the next event activation time (called **wakeUp-Time**) and sends a **wakeUp** message to the corresponding object. The receiver can respond to it in its own unique manner. In addition, the simulation model will also require a 'coordinator' object that will be informed of all important events. We call this class **Coordinator**. Note that we are using names from BLOCS/M (Glassey and Adiga, 1989; 1990).

In order to incorporate persistence, each object must know how to save itself in the database as well as recreate itself. This means that each object must implement **save, destroy** and **read** methods. In this manner Tracking simply has to simply send the appropriate message to the object when it wants a change recorded in the database.

5.6.3 Step 3. Entity–relationship (ER) model

To keep the ER diagram down to a manageable size, we will model the system in sections. The sections are:

1. resources and queues,
2. operations and operation sequences,
3. resource allocation; and
4. decision making.

Resources and queues

Consider one of the workstations in the facility. It is a physical object and is a candidate for representation as an entity. The information that is associated with any workstation is (following our description and assumptions): the name of the workstation and the set of identical machines that belong to it. Each machine is a physical object and hence can be an entity. The attributes that can be associated with this entity are: the product it is set up for and the state it is in (whether **BUSY, IDLE** or **INSETUP**). Hence, for each combination of the two values we will have a distinct machine. Any of these machines is an entity. All such machines can be grouped into an entity set called **Machine** which has the attributes: **setupName** and **state**. In the same way we have an entity set called **WorkStation** with the attributes: **name** and **machineList**.

We can have another entity set called **Lot** with its own attributes: **name** and **noOfWafers**. Each workstation (usually) has a queue of lots awaiting processing. The only attribute the queue can have (for our simple model) is **noInQueue**, the number of lots in queue. The ERD for this section on resources and queues is shown in Figure 5.4.

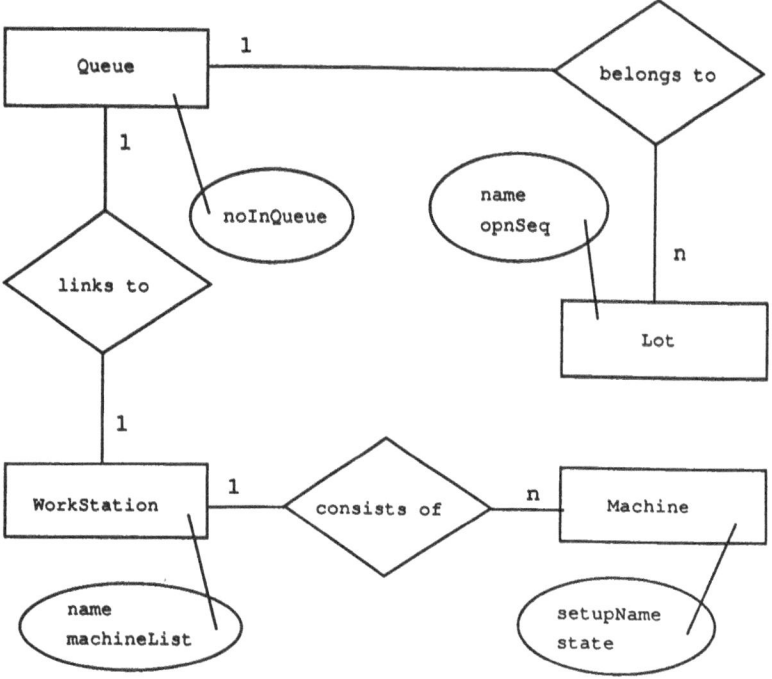

Figure 5.4 *ERD showing the resource and queue relationships.*

We note that there is a one-to-one relationship between a workstation and its associated queue. (Incidentally, this is an assumption made in this model.) While workstations are physical entities that need to be represented as entities, perhaps we do not need an entity set called **Queue**, so **noInQueue** could be an attribute of **WorkStation**. This dilemma can be resolved by asking whether a queue can exist as an independent entity. The answer is 'yes', we can have a queue of finished products awaiting shipment that is not associated with any workstation. If we were considering machine failures, we could have a queue of workstations awaiting repair. Hence we conclude that **Queue** is an independent entity set. We can name the queues in order to distinguish them. Thus the attributes of **Queue** are: **name** and **noInQueue**.

We find that while **Queue** and **WorkStation** are entity sets, either can exist independently of the other. The membership in the relationship **linkedTo** is one-to-one and non-mandatory for both types. Following logical data modeling guidelines (Teorey *et al.*, 1986) for such a relationship, we need another data structure that will represent the link between workstations and the queues associated with them.

Operations and operation sequences

Lots are processed (tested) at the workstations. Each lot is associated with an 'operation sequence' which lists the sequence of operations the lot has to undergo. The set of operation sequences map into an entity set **OpnSeq**. The relationship between **Lot** and **OpnSeq** can be represented in the ER approach by the descriptor **follows**. An operation can be performed at any one of a set of different workstations. The set of operation entities map into an entity set called **Operation**. The ER diagram for these relationships is shown in Figure 5.5.

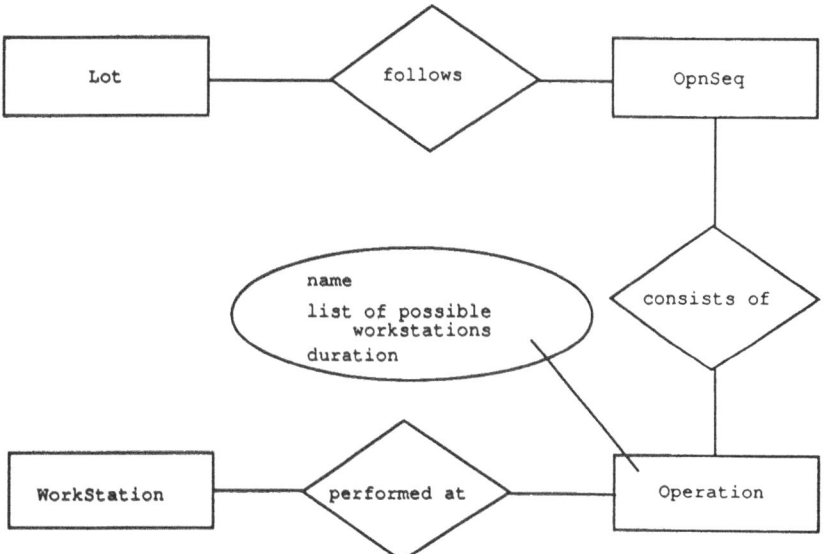

Figure 5.5 *ERD showing the operation sequence relationships.*

Resource allocation

A resource allocation decision is made by the coordinator whenever a change in the state of the test area model occurs. The decision lasts for a particular duration. For example, a change of state occurs when a lot finishes processing at a workstation. (The four possible state changes are listed above.) For our model, an allocation of resources could be for the processing of a lot of wafers or for performing a setup at a workstation. Note that while the decision process results in an allocation of resources, the decision process itself is independent of the actual resource allocation.

The relationship between the resources, coordinator and the timer for the processing of a lot is shown in Figure 5.6. This is a 4-ary relationship and can be redrawn using established guidelines (Hawryszkiewycz, 1984; p.127) to pro-

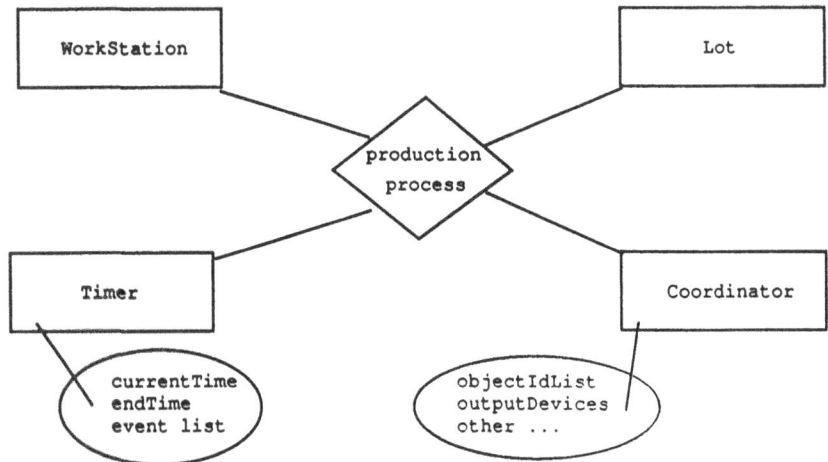

Figure 5.6 *ERD showing the allocation of resources for production.*

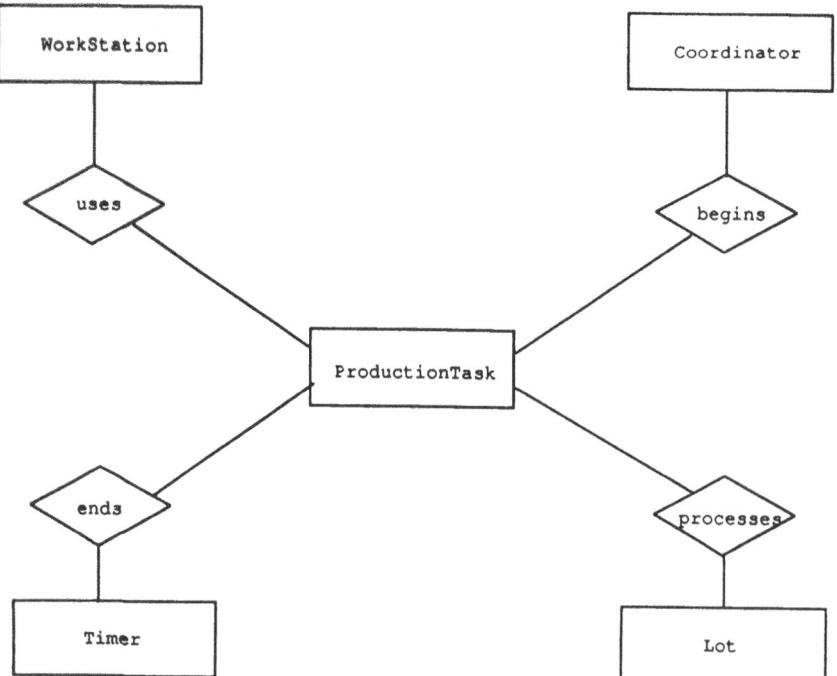

Figure 5.7 *Remodeling the resource allocation relationships at production.*

duce Figure 5.7. Similarly, we can draw the relationships that exist when a set-up is to be performed (Figure 5.8) and redraw it to produce Figure 5.9.

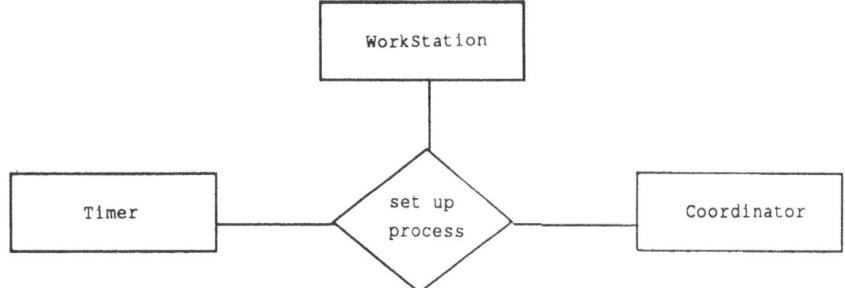

Figure 5.8 *ERD showing the allocation of resources for setup.*

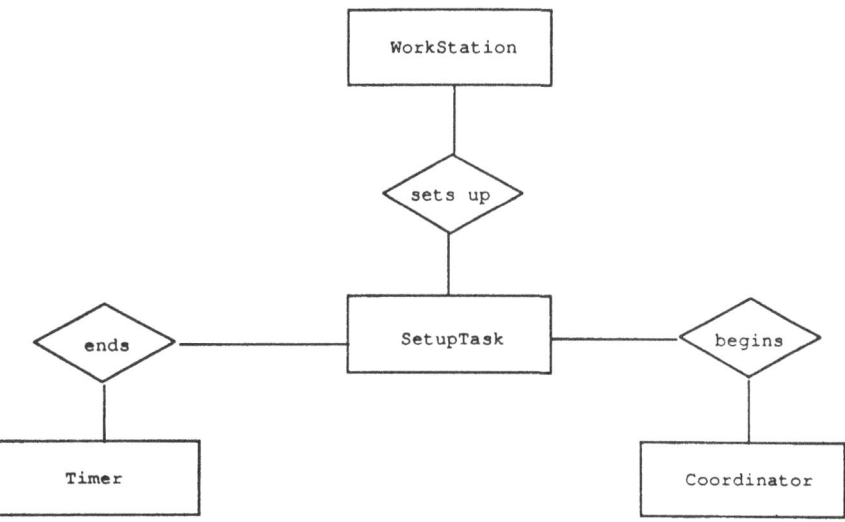

Figure 5.9 *Remodeling the resource allocation relationships for setup.*

Decision-making

Decisions have to be made (in order to decide) whether to allocate resources when a state change occurs and if so, which resources to allocate. For example, when a workstation becomes idle, we have to decide which lot to process next. Heuristics and queue disciplines are used to make such decisions. Usually, the kind of heuristics used depends on the workstation where production has to be scheduled. For example, at a non-critical workstation we may simply use first-in-first-out (FIFO) logic. At a bottleneck workstation however, we would like to minimize setup changes and hence FIFO may not be useful. Some other dispatching heuristic may be required.

To test different heuristics and policies conveniently, we would like a clear separation between the decision logic and the system it acts on. We will need an entity set that simply contains the scheduling heuristics, which we call **DInfo** (short for 'decision information'). **DInfo** will consists of heuristics (for example FIFO or 'starvation avoidance'). The ER diagram for the relationships that exist in the context of decision making is presented in Figure 5.10.

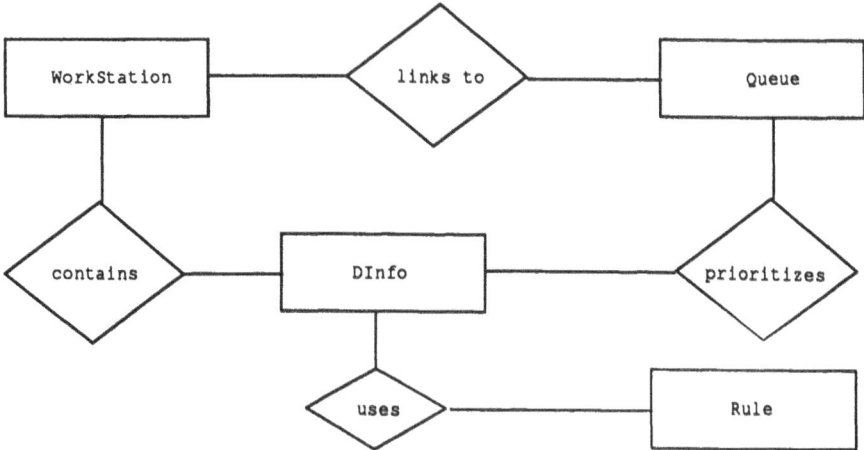

Figure 5.10   *The DInfo entity set.*

5.6.4 Step 4: Identify classes and instance variables from the ERD

The entity set **WorkStation** maps into an object class **WorkStation**. A specific work station with properties: name = 'Stepper' and **machineList** is an entity and will be represented by an instance of class **WorkStation**.

From Figure 5.5 we conclude that **Operation** must be an object class. From Figures 5.7 and 5.9 we obtain new classes **ProductionTask** and **SetupTask**. The attributes of these entity sets while as yet unknown, will become obvious from the analysis of the message flow diagrams.

A partial list of objects identified thus far and their instance variables follows. We have used the naming conventions followed in BLOCS/M (Glassey and Adiga, 1990).

1. **Timer**
   (a) **currentTime**;
   (b) **endTime** (time the simulation should stop);
   (c) list of future events ordered by their activation times.
2. **Machine**
   (a) **state** (BUSY/IDLE/INSETUP);
   (b) **setupName**.
3. **WorkStation**
   (a) **name**;

(b) **machineList** (an ordered collection of the machines that belong to it);

(c) **id** of **DInfo** object it is associated with.

4. **Queue**

(a) **name** (for example, 'FinishedLots');

(b) ordered collection of **Lots** that belong to it.

5. **Lot**

(a) name;

(b) **state (WAITING/INPROCESS)**;

(c) **id** of **OpnSeq** it is following.

6. **DInfo**

(a) name of rule that it represents (for example, FIFO, SRPT, etc.).

5.6.5 Step 5: Identify important events and the objects involved

In the case of a production system, an event of interest could be the start of production at a particular workstation. The objects involved are the workstation, the lot to be processed and the operator in charge of the workstation. At the moment of starting production, the operator may load the workstation and start processing.

The following are the events of interest for the facility:

1. start production (testing) on a particular lot at a particular workstation;
2. end production (testing) on a particular lot at a particular workstation;
3. start performing a setup on a workstation;
4. end performing a setup on a workstation;
5. receipt of a new order by costing.

Let us analyse the production steps. Starting production is an action taken by **Coordinator** and takes the form of assigning resources (workstations, lots) to perform a task (i.e. a production operation). When we start production, it is for a lot at a particular workstation. It lasts for a particular duration; once the operation is finished, the lot is removed from the workstation and added to the next queue. The workstation goes idle. It is time for **Coordinator** to make the next dispatching decision.

5.6.6 Step 6: Draw MFDs for each event and identify methods

Let us continue analyzing the production events. At the start of production, **ProductionTask** sends messages as shown in the message flow diagram in Figure 5.11. The first message is from **Coordinator** creating an instance of **ProductionTask** for a particular instance of **Lot** (let us call this **aLot**) at a particular instance of **WorkStation** (let us call this **aWkStn**). The duration of the processing step is stored in the instance of **Operation** that is to be performed. We have access to this instance of **Operation** via the **OpnSeq** that **aLot** is linked to. Hence **ProductionTask** queries **aLot** for the duration. The object **aLot** in turn queries its **OpnSeq**, gets the information and returns the processing duration for the step to **ProductionTask**. Then the newly created instance of **ProductionTask** sends **startProduction** messages to **aWkStn**, **aLot** and **Coordinator**. These respond by altering their respective states (for example **aLot** will go from **WAITING** to **INPROCESS**).

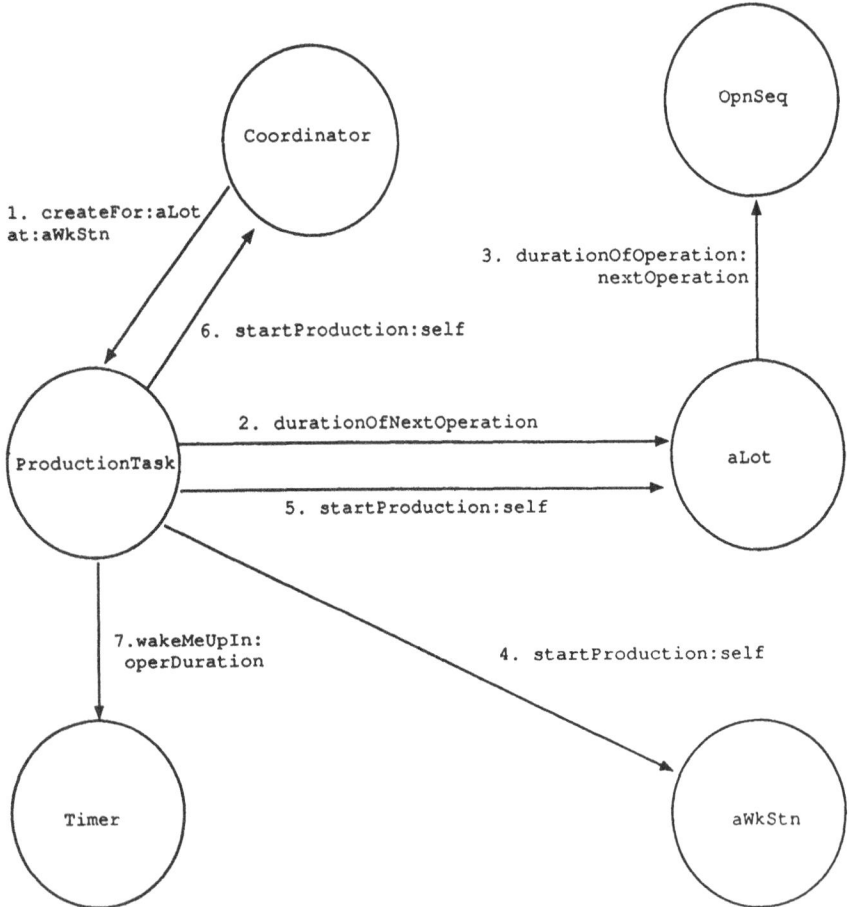

Figure 5.11 *MFD for ProductionTask at the start of production.*

The next message **wakeMeUpIn:opnDuration** is to **Timer**. **Timer** responds by adding the object to its event calendar. Its position in this calendar is according to its **wakeUpTime** which is calculated as follows:

$$\text{wakeUpTime} = \text{currentTime} + \text{opnDuration}$$

When the **wakeUpTime** for this object is reached, **Timer** will send it a **wakeUp** message. The **ProductionTask** instance will respond by sending the **endProduction** message to **aWkStn** and **aLot**. These will alter their states appropriately.

An **endProduction** message to **Coordinator** is also sent. Figure 5.12 shows the message flows triggered by the end of the production event.

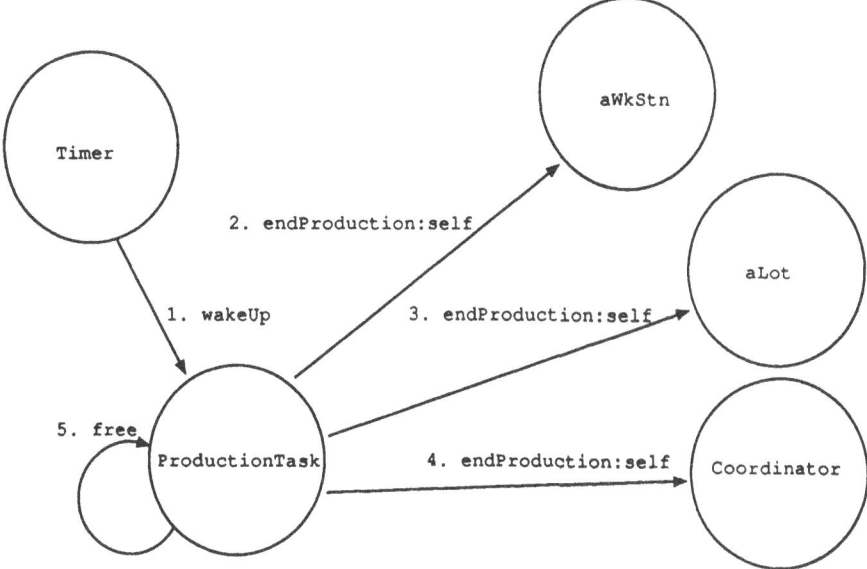

Figure 5.12 *MFD for ProductionTask at the end of production.*

Coordinator will respond by writing to statistical files and moving the lot to the next queue. According to user definitions, other functions may be executed. The **SetupTask** will behave similarly. Objects identified in this step and their instance variables are:

1. **ProductionTask**
   (a) **id** of the **Lot** to be processed;
   (b) **id** of the **WorkStation** to be used;
   (c) operation duration.
2. **SetupTask**
   (a) setup name;
   (b) setup duration;
   (c) **id** of **WorkStation** of which the machine to be set up is a part.

Once a task is finished it need not be tracked any longer. Hence an instance of **ProductionTask** or **SetupTask** is a temporary object. Once finished it may be freed from memory.

Identify hierarchies

We could apply the *generalization* construct to identify class hierarchies in this situation. We can say that **ProductionTask** *is-a* **Task**. Also **SetupTask** *is-a* **Task**. In this manner **Task** is identified as a superclass to **ProductionTask** and

SetupTask. One of its instance variables would be **taskDuration**. It would also implement a method called **wakeUp**. This method would be more like a place holder. The subclasses will implement the actual functionality. Part of its functionality would be to simply send a message to **Coordinator** and free itself.

1. Task
   (a) taskDuration

In future extensions to our model we may wish to consider events such as random breakdowns. In the context of discrete event simulations, such events will be generated and added to **Timer**. In such situations, we could state 'Task *is-a* Future-Event'. In this manner the hierarchy could grow with the complexity of the model.

5.6.7 Step 7: Revise and consolidate object specifications

We identified the instance variables required to support the data needs of the objects in Step 4. Now we shall look at the methods necessary for the objects to respond to messages representing the interactions between objects participating in important events. From Figure 5.11 we get the following list of methods. A message maps into a method on the interface of the receiver object. Below each method we list the functionality to be provided by this object.

1. **Coordinator** to **ProductionTask** – **createFor:aLot at:aWkStn**
   Note that the newly created instance of **ProductionTask** identifies itself to the objects to which it sends the **startProduction** message. This is done by passing the **self** variable (Stepstone, 1989).
   (a) set instance variables to **aWkStn** and **aLot**;
   (b) obtain operation time from **aLot**;
   (c) send **startProduction**: self message to **aWkStn** and **aLot**;
   (d) **wakeMeUpIn:opnDuration** message to **Timer**.
2. **ProductionTask** to **aLot** – **durationOfNextOperation**
   (a) query associated **OpnSeq** for processing duration of next step – this indicates that **Lot** must have another instance variable **nextOperation** to keep track of the operation it is to undergo next;
   (b) return value.
3. **ProductionTask** to **aLot** – **startProduction:self**
   (a) change **state** to **INPROCESS**.
4. **ProductionTask** to **aWkStn** – **startProduction:self**
   (a) obtain **id** of **aLot** (the lot to be processed) from the instance of **ProductionTask**;
   (b) get the **id** of the operation that is to be performed from **aLot**;
   (c) ensure that there exists at least one machine (out of **machineList**) that has both the correct setup and is in state **IDLE**;
   (d) if both conditions hold, send **startProduction:aTask** message to that machine, else, return error messages.
5. **ProductionTask** to **Timer** – **wakeMeUpIn: opnDuration**

(a) compute **wakeUpTime** = **currentTime** + **opnDuration**;

(b) add **ProductionTask** to event stack in order sorted by **wakeUpTime**.

6. **Lot** to **OpnSeq** – durationOfOperation: **nextOperation**

(a) return processing duration for specified operation.

From Figure 5.12 we get the following specifications for the messages:

1. **Timer** to **ProductionTask** – **wakeUp**. Similar to start of production, the instance of **ProductionTask** that has been woken up identifies itself to the objects to which it sends the **endProduction** messages by passing the **self** variable (Stepstone, 1989).

(a) send **endProduction:self** messages to **aWkStn**, **aLot** and **Coordinator**;

(b) free **self**.

2. **ProductionTask** to **aLot** – **endProduction:self**

(a) change state to **WAITING**.

3. **ProductionTask** to **aWkStn** – **endProduction:self**

(a) send **endProduction:aTask** message to all machines in **machineList**.

(i) **WorkStation** to **Machine** – endProduction: **aTask**. If the machine is currently being worked on by **aTask**, then it will respond by changing **state** from **BUSY** to **IDLE**. (Note that we have not discussed **Machine** in detail for this example.)

4. **ProductionTask** to **Coordinator** – **endProduction:self**

(a) query the appropriate **DInfo** to determine the next action and implement it;

(b) any other user requirements.

It is important to note that Steps 2 through 7 may have to be iterated in order to arrive at an acceptable design.

We now summarize the information derived in the previous steps in Table 5.1. Messages appear as methods in the receiver objects. The information in Table 5.1 can be combined with the instance variable information from the preceding steps to obtain specification sheets for each class. This is illustrated in Table 5.2 for two classes: **WorkStation** and **ProductionTask**.

Table 5.1 A partial summary of message flows

No.	SenderObject	ReceiverObject	Message
1	Coordination	ProductionTask	createFor:aLot at:aWkStn
2	Timer	ProductionTask	wakeUp
3	ProductionTask	Lot	durationOfNextOperation
			startProduction:self
			endProduction:self
		WorkStation	startProduction:self
			endProduction:self
		Timer	wakeMeUpIn:opnDuration
		Coordinator	startProduction:self
			endProduction:self
4	Lot	OpnSeq	durationOfOperation: nextOperation

Table 5.2 A partial list of object specifications

No.	Object name	Instance variables	Methods
1	WorkStation	name	createName: noOfMachines:(int)
		machineList	startProduction:
			startSetup:
			endProduction:
			endSetup:
			save, destroy, read
2	ProductionTask	lot	createFor: at:
		WorkStation	wakeUp
		opnDuration	save, destroy, read

5.6.8 Step 8: Implement the design in an OOP language

We assume that the developer has already made a choice as to the implementation OOPL. Discussing OOPLs and criteria for selecting one is beyond the scope of this chapter. A review of OOPLs with special emphasis on C++ (Ellis and Stroustrup, 1990) and Objective-C is presented in Chapter 7. In this example we implement the specifications in Objective-C (Stepstone, 1989). For the sake of brevity, we limit ourselves to the interface and implementation files for only **ProductionTask**. The code is presented in the Appendix to this chapter. Readers not familiar with the syntax of Objective-C can read Chapter 7 before reading the Appendix.

5.6.9 Using the object library in other modules (subsystems)

Thus far, the object library has been developed mainly in the context of Production Management and simulation. However, the classes can be easily reused in the context of say, Tracking. When the **endProduction** message from Process Control is received (see Figure 5.3), The Tracking Coordinator can set the new values on the appropriate **WorkStation** and **Lot** objects and send the **save** message to them.

When **ProductionTask** sends the **wakeMeUpIn** message to **Timer** at the start of production (see Figure 5.11) the time implemented for Tracking has to send the **save** method to the instance of **ProductionTask** instead of adding it to its list of future events. The actual end of production signal will come from Process Control.

We find that our method has led to a domain-specific object library that can be easily reused in different contexts with only minor, localized modifications. Generally, these modifications result in new instance variables and methods in existing objects or the addition of new object classes.

5.6.10 Documentation

As we have mentioned above, timely documentation is essential for deriving the full benefits of object-oriented programming. We believe that all relevant documentation should be part of the source files. If this is incorporated in a standard format, then text processing utilities can automatically generate specification sheets from the source files.

We present a format for documentation in the Appendix. It is incorporated into the source files for the example **ProductionTask** class. This is a minimal set that users can modify and build on for their own specific requirements. It is based on the documentation standards used for the BLOCS/M Library and is somewhat tuned to the Objective-C programming language. Users may modify these guidelines where appropriate for their own environments and needs.

5.7 State transition diagrams

In our method we have avoided incorporating (and thus mandating) the various types of diagrams that can be drawn. The method is minimal and simple. It does not attempt to impose on the developer any unnecessary types of analyses. However, in certain cases, augmenting the method with other simple techniques can actually serve to enhance the clarity of the design. One such technique is the state transition diagram.

The behavior of finite state entities and processes can be illustrated via a state transition diagram. There are two ways in which such diagrams are drawn, bubble form (also called Moore diagrams) and rectangular form (Kowal, 1988).

Analyses of state transitions and that of events are closely tied together. Either can be performed. Part of the information derived from an event analysis relates to the state changes that the interacting entities undergo. Through state transition analyses we can identify the events that trigger the state changes. It is a cyclical dependency.

In fact, a method such as ours could conceivably have used either technique. However, it is easier for the end users as well as developers to view and conceptualize the system as receiving and responding to events. For example, the user (or even the developer) will find it much easier to understand the behavior of a graphical user interface (GUI) application as responding to events such as a mouse click or key press, rather than from the viewpoint of a window in the GUI going from State-A to State-B. Also, while state transition diagrams are useful for finite state processes, they can be hard to deal with when the states are not finite or when all possible states are not immediately obvious. Hence, an event analysis is preferable to a state transition analysis.

However, augmenting an event analysis with a state transition analysis can serve to better one's understanding of system behavior, especially for large and complex systems. It can also serve to identify any omissions in our event analyses, such as identifying illegal state transitions.

We are reluctant to burden designers and developers with yet another requirement or technique. However, we recommend that state transition diagrams be used not as a mandate but as an aid to understanding the behavior of more complex entities and processes. A state transition diagram for a workstation object in our example is shown in the Figure 5.13. This diagram is useful in illustrating the fact that a workstation cannot go from a **BUSY** state to an **INSE-TUP** state.

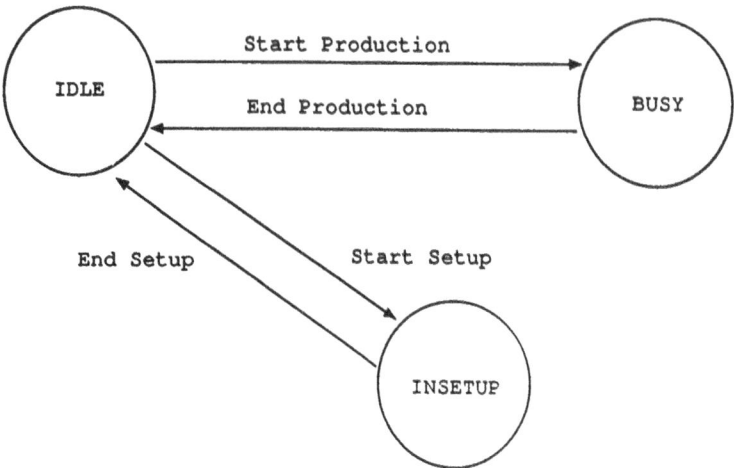

Figure 5.13 *State transition diagram for WorkStation.*

5.8 Conclusion

While different design methods have been presented for the object-oriented decomposition of software, none applies the object-oriented paradigm consistently from the system level all the way down to the code level. No method supports techniques for capturing object-oriented features such as messaging and polymorphism. Few methods support rapid prototyping, an essential activity for developing reusable components and accelerating the learning curve. And none addresses the need for the rapid development of domain-specific, reusable object libraries. In this chapter, we have proposed a model-based prototyping approach that addresses the above weaknesses in the existing approaches. We have described the approach through an example. Such a model-based approach should be easier to understand and is, probably, more maintainable than the conventional approaches. One reason for this assumption is that a model-based approach reflects the user's view of reality. It should be easier to relate changes that may take place in the user's environment to object requirements, as real-world events are simulated in the design process.

One may perceive a weakness in our approach. It does not belong to a 'cookbook' class of approaches. A cookbook approach may be characterized by a list of forms to be filled, a comprehensive notation scheme, and volumes of procedures to be followed. The strength of such approaches is that analysis and design is reduced to a set of simple, discrete activities and steps. This allows people with minimal training and experience to be productive in writing software. A strict adherence to the approach is expected to result in a feasible solution, i.e. usable software that meets the requirements.

We acknowledge that certain steps have to be followed in designing software but we do believe that an iteration process is required to refine the engineering of object-oriented software. Further, it is not our aim to propose a 'cookbook' approach to design. We have presented an approach that we believe is flexible, conceptually sound and allows one to prototype OOS by taking advantage of many attractive features of object-orientation.

A potential weakness of an event-based analysis is that one may not be able to take advantage of commonality among events, thus limiting reusability of code. We believe that a data modeling effort often brings out important common associations between objects affected by events as objects. An example of this is the **Task** object class in the example.

While this approach is simple yet rich enough to help prototyping applications, we do acknowledge that a CASE (Computer-aided software engineering) tool may make life much easier for larger projects! The ideas presented in this chapter can be conveniently represented by extending CASE tools that use the entity – relationship approach.

5.9 References

Adiga, S. and Gadre, M. (1990) Object-oriented software modeling of a flexible manufacturing system. *Journal of Intelligent and Robotic Systems*, **3** 147–165.

Beck, K. and Cunningham, W. (1989) A laboratory for teaching object-oriented thinking. *SIG-PLAN Notices*, **24** (10).

Boehm, B. W. (1988) A spiral model for software development and enhancement. *IEEE Computer*, **21**, (5).

Booch, G. (1986) Object-oriented development. *IEEE Transactions on Software Engineering*, SE 12:2, February, 211–21.

Booch, G. (1991) *Object-oriented design with applications.* Benjamin/Cummings Publishing Co. Inc.

Chen, P. P. (1976) The entity–relationship model – toward a unified view of data. *ACM Transactions on Database Systems.* 9–36.

Coad, P. and Yourdon, E. (1990) *Object-oriented analysis.* Yourdon Press.

Date, C. J. (1990) *Introduction to database systems*, Fifth edition. Addison-Wesley. Reading, MA.

Ellis, M. A. and Stroustrup, B. (1990) *The annotated C++ reference manual.* Addison-Wesley, Reading, MA.

Glassey, C. R. and Adiga, S. (1989) Conceptual design of a software object library for simulation of semiconductor manufacturing systems. *Journal of Object-Oriented Programming*, November/December, 39–43.

Glassey, C. R. and Adiga, S. (1990) Berkeley Library of Objects for Control and Simulation of Manufacturing (BLOCS/M). *Applications of Object-Oriented Programming*, L. Pinson and R. Wiener (eds.), Addison-Wesley, Reading, MA. **2**. (4).

Hawryszkiewycz, I. T. (1984) *Database analysis and design.* Science Research Associates Inc.

Henderson-Sellers, B. and Edwards, J. M. (1990) The object-oriented systems life cycle. *Communications of the ACM*, **33**(9), 142–159.

Kowal, J. A. (1988) *Analyzing systems.* Prentice Hall.

Law, A. M. and Kelton, W. D. (1982) *Simulation modeling and analysis.* McGraw-Hill.

Lawson, H. W. (1990) Philosophies for engineering computer-based systems. *IEEE Computer*, **23**(12), 52–63.

Ormsby, A. (1991) Object-oriented design methods. *Object-oriented languages, systems and applications*, G. Blair, J. Gallagher, D. Hutchison and D. Shepherd (eds.), Halsted Press.

Ozkarahan, E. (1990) *Database management – concepts, design and practice*, Prentice-Hall.

Potter, W. D. and Trueblood, R. P. (1988) Traditional, semantic and hyper-semantic approaches to data modeling. *IEEE Computer*, **21**(6), 53–63.

Power, L. (1987) Specification and design of objects. *OOPSLA '87 – Addendum to the Proceedings*, October, ACM Press, NY.

Rumbaugh, J., Blaha, M., Premerlani, W., Eddy, F. and Lorensen, W. (1991) *Object-oriented modeling and design*, Prentice-Hall.

Shaw, C. (1987) The open channel – build a model of the application world before you begin designing your database application system. *COMPUTER*, July, 104–5.

Shlaer, S. and Mellor, S. J. (1988) *Object-oriented systems analysis – modeling the world in data.* Yourdon Press.

Stepstone Corporation (1989) *The Objective-C Compiler v4.0 and ICpak 101 v4.0*, Stepstone Corporation, Sandy Hook, CT.

Teorey, T. J., Yang, D. and Fry, J. P. (1986) A logical design methodology for relational databases. *ACM Computing Surveys*, **18**(2), 197–222.

Thomas, D. (1989) In search of an object-oriented development process. *Journal of Object-Oriented Programming*, **2**(1), 60–3.

Winblad, A. L., Edwards, S. D. and King, D. R. (1990) *Object-Oriented Software.* Addison-Wesley Reading, MA.

Wirfs-Brock, R. J. and Johnson, R. E. (1990) Surveying current research in object-oriented design. *Communications of the ACM*, **33**, (9), 104–124.

Appendix

We present Objective-C code for the interface and implementation of class **ProductionTask**. For specific details on syntax and other language related matters, users are referred to the appropriate manuals (Stepstone, 1989). In Objective-C the class definition is composed of two files: the interface file and the implementation file. The interface file has the name of the class with a '.h' extension. The implementation file has the name of the class with a '.m' extension. The naming conventions and programming style followed are from BLOCS/M (Glassey and Adiga, 1990).

Class names begin with an upper case letter. Objects (which are instances of a class) have names beginning with a lower case letter. For example the class is **Machine** and the instance is **aMachine** or **machine**. The message syntax is

[**receiverObject messageName**: (an optional set of arguments)];

Since we have the benefits of inheritance, if an instance variable and/or method is not found in the receiver object, Objective-C will search its super classes for it. Hence if the reader finds that a method or instance variable being used is not defined in the following lines of code, he or she may assume that it exists in one of the superclasses.

timer	Global instance of class Timer.
coordinator	Global instance of class Coordinator.
Object	The class at the root of the inheritance tree. Defined in Objective-C.
id	New variable type defined in Objective-C for pointers to objects.
self	Type **id** reserved name that points to the object we are working on.
super	Type **id** reserved name that points to immediate super class.
//	Signifies that the rest of the line is for comments.

A.1 A documentation template

```
//        ==== [INTERFACE FILE : ProductionTask.h] ====
//
//    Class Name              ProductionTask
//
//    Superclass              Task
//
//    Source Files            ProductionTask.h
//                            ProductionTask.m
//
//    Other classes and Files Used    Timer.h
//                            stdio.h
//                            math.h
//
//    Global Instances        None
//
//    Author                  Milind Gadre      (01/15/91)
//
//    Revised By              S. Adiga          (05/02/91)
//
//    Introduction : ProductionTask models the interaction of physical
//    elements such as machines and operators during the activity of
//    production. It also encapsulates the inter-object communication
//    at the start and end of this activity. An instance of this class
//    is created at the start of production. This instance is dormant
```

```
//     until triggered at the end of production (either by a
//     real-world or simulation event).
//

#import 'Task.h'

@interface ProductionTask : Task
{
       //     === [Instance Variables] ===

       // Note that taskDuration is inherited from the Task superclass.

       id    lot,       // the lot to be processed
             wkStn;     // the id of the workstation at which the lot
                        // is to be processed
}
//            ==== [Factory Method(s)] ====

// This method is used to create an instance of ProductionTask for a
// specified lot at a specified workstation. This method will query the
// lot for the operation duration for its current step. Then it will send
// startProduction messages to the lot, workstation and coordinator. Finally
// it will send the wakeMeUpIn message to timer.

+ createFor: (id)aLot at: (id)aWkStn;

//            ==== [Instance Method(s)] ====

// This method implements the logic for the end of production. First it
// sends endProduction messages to the lot, workstation and coordinator.
// Then it frees itself (from memory).

- wakeUp;

// The following methods simply return the associated instance variables.

- wkStn;

- lot;

// The following methods are used to communicate with the database. Any
// error value is returned as a long integer.

- (long) save;

- (long) destroy;

- (long) read;

@end       // end of class interface

//
//            ==== [IMPLEMENTATION FILE : ProductionTask.m] ====
//

extern id  timer,       // global instance of class Timer
           coordinator; // global instance of class Coordinator
```

```objc
#import 'ProductionTask.h'

@implementation ProductionTask


+ createFor: (id)aLot at: (id) aWkStn
{
    self           = [self new];
    wkStn          = aWkStn;
    lot            = aLot;
    taskDuration   = [lot durationOfNextOperation];

    [wkStn         startProduction:self];
    [lot           startProduction:self];
    [coordinator   startProduction:self];
    [timer         wakeMeUpIn:taskDuration];

    return self;
}
- wakeUp
{
    [wkStn         endProduction:self];
    [lot           endProduction:self];
    [coordinator   endProduction:self];

    [self free];
    return self;

}

- wkStn                      {return wkStn;}

-.lot                        {return lot;}

- (long) save
{
    // Implement communication with DB
    // Return any errors.
}

- (long) destroy
{
    // Implement communication with DB
    // Return any errors.
}

- (long) read
{
    // Implement communication with DB
    // Return any errors.
}

@end      // end of class implementation
```

6 Object-oriented databases

S. ADIGA and J. KOLYER

Database management systems (DBMS) emerged as the need to process large amounts of data became important in the database industry. Early attempts resulted in 'file systems' that provided sequential access to data stored as records in files. To accommodate various needs of different programs, multiple copies of the same data were made and sorted in various orders (Loomis, 1990). Collections of programs for sorting, indexing etc. made up the 'file system'. Hierarchical and Network databases evolved because of the needs of data storage to be accessed by multiple programs in multiple ways. These requirements also gave rise to mechanisms for concurrency control, recovery etc., which in turn gave rise to the idea of database management systems (DBMS).

Relational DBMS represents the third generation systems in data management. A relational DBMS represents data in tables, with columns representing the properties of the tables and rows the values of those properties. This data model gained popularity in the 1980s due to its simplicity, and superior query facilities based on their algebraic properties. An additional advantage is that an application programmer is not required to know anything about the physical storage of the data being manipulated. The relational DBMS are discussed in Date (1990), for example.

Object-Oriented Databases (OODBs) have grown from both the addition of semantic constructs to the relational model and specialized needs of select database applications. An extensive list of semantic constructs that add to the relational model is seen in Stonebraker (1991). The research areas feeding the development of OODBs have been object-oriented languages, programming with persistent data structures, next generation databases, and semantic data models. Also, the needs of certain applications have not been addressed satisfactorily with the relational model. These niche applications are primarily computer-aided design (CAD), computer-aided software engineering (CASE), office information systems (OIS) (including multimedia database systems), and artificial intelligence (AI).

The topic of OODB has been discussed in much detail as seen in the books by Kim and Lochovsky (1989), Gupta and Horowitz (1990) or Zdonik and Maier (1990). In this chapter, we will consider select topics that impact the standing of OODBMS relative to relational DBMS. We have chosen these topics on the basis of their relevance to two critical components of an OODB system, i.e. usability and performance. The special needs of some engineering applications are discussed in section 1. Section 2 will outline some issues regarding an object oriented data model – an unresolved factor, yet critical to acceptance of OODBs. Sections 3, 4 and 5 will discuss object identity, schema evolution and versioning respectively. The latter two affect usability, particularly important for its niche applications. Section 6 discusses storage considerations, including object persistence, which is a performance criterion. Section 7 will consider the issue of querying, both on performance and usability criteria; we also discuss complications involved with optimization and indexing.

6.1 Needs of the engineering/manufacturing domains

Two examples of typical engineering applications where an object-oriented database (OODB) may be useful are CAD (computer-aided design) and CASE (computer-aided software engineering) (Cattell and Skeen, 1990). We also briefly discuss its relation to manufacturing and concurrent engineering.

6.1.1 CAD and CASE

In a CAD application, such as the design of a printed circuit board, components and their interconnections might be stored in a database; an optimization algorithm may need to follow connections between components and rearrange them to reduce wire lengths.

In a CASE application, program modules and their interconnections might be stored in a database; a system-building algorithm may need to traverse the dependency graph examining version numbers to construct a compilation plan.

In both cases, hundreds or thousands of objects are accessed per second (equivalent of a relational join for each object). OODBs provide faster and more convenient support to these activities as objects are accessed directly. Increased modeling power allows objects to be modeled very close to reality, reducing or eliminating the joins required (Gupta *et al.*, 1989).

Large design or software engineering projects usually require many professionals to work together. A design environment should be able to support cooperative work by a networked group of engineers. Engineers must be able to create new versions of existing parts without overwriting someone else's work, or work on his/her own design without being preempted by others. OODBMS support such collaborative work by allowing team members access to others' work-in-progress. Users can check out an object or group of objects at the same time as another user, saving their changes or merging them into the same version of the object.

6.1.2 Storing manufacturing-related data

While RDBMS essentially search through rows and columns of data, OODBMS add the ability to operate on large files that may include voice, graphics, text or complex data structures such as arrays and lists. An OODBMS can be used to bring up not only a part number and its description, but also the drawings or image of the part to be ordered or produced as the case may be.

Retrieving data such as a product's bill of materials from a relational database can be quite a time-consuming process. The system must search the database each time it has to trace a link between a subcomponent and the next one. An OODB can store the product as a composite object with the links to its components represented as direct references (id's). Thus the OODBMS can go directly to the object instead of searching through tables to find matching values. This will allow it to retrieve a bill of material in one query, in a fraction of the time required for a relational system.

6.1.3 CAD and the manufacturing link

The basic concept of encapsulation of data and methods in an object offers an advantage in promoting an integrated relationship between CAD and manufacturing. It allows us to build relationships in a convenient manner. For example, if the diameter of a bolt changes, an OODB should be able to propagate changes to, or notify the designer of the need for change to the nut corresponding to the bolt. Such facilities are quite useful in building applications such as 'design for manufacture'.

This connection between design and manufacturing can be exploited to a great advantage in the area of 'concurrent engineering'. Concurrent engineering is a recent research thrust towards a team effort to product design. An important activity in this field is to consider the 'downstream' effects of product design on manufacturing, marketing and support. An OODB can provide the modeling power for representing complex objects and also facilities for instituting systems controls sharing data, broadcasting change, etc.

6.2 Towards an object model for database management

OODBs appear to have evolved through construction between two extremes. These range from an object-oriented programming language (OOPL) with extensions for persistent storage, to a fully-fledged DBMS utilizing concepts derived from the OOPLs. System implementations have preceded a coherent, formal object data model. To date, there is still no accepted foundational model for OODB systems (OODBS), but there are quite a few commercially-available systems (e.g. Versant, ObjectStore, GemStone, etc.). This, in part, reflects that OODBS are chiefly of a market-driven concept, borrowing principles that work well in other applications such as software engineering. Following the success of these principles, and the inadequacy of current data models for the niche applications, there has ensued an attempt at integrating diverse concepts and methodologies under the 'object-oriented' umbrella.

Another reason for the lack of an object data model is that the classical notion of a data model is inappropriate for the object-based concepts. There are fundamental philosophical inconsistencies between the two (as described below). Nevertheless, a formal framework is possible, if not in the classical sense. A framework is also necessary to provide a foundation for further development of OODBs and continued theoretical investigation. Beyond this, there is disagreement among database professionals about whether to develop a whole new data model or to simply extend current models to do the same job.

The degree to which OODBs gain acceptance commercially and within the database community will determine their success. The commercial support will ensure continued development on implementation, while support in the database community will provide continued theoretical development. The absence of a formal model affects the chances of acceptance, particularly in the database community. Without this formalization, a system's integrity and robustness cannot be guaranteed. In part, what a formal model should provide are consistency and completeness. Roughly speaking, consistency ensures that the system is not self-contradictory, i.e. contradicting an established set of principles or rules. Completeness ensures that the set of all possible (or at least desirable) actions can be achieved within the framework.

6.2.1 Object-orientation in databases

Conceptually, object orientation and databases make incompatible demands upon a system. The object-oriented approach emphasizes object independence through encapsulation, communicating with the outside world through a well-defined interface via message passing. Hence objects are independent of each other, and from this arises much of the power of the OO approach. Yet databases traditionally emphasize data independence, separating the world into two parts: data and applications. This ensures a well-defined responsibility of the database system, which simplifies the system interface. Upon blending these two philosophies, we have objects containing data, and objects acting as the components of the application. This inconsistency provides ambiguity, and it is the fundamental difference between the object-oriented paradigm and traditional database systems.

Many OODBS incorporate two methodologies for viewing objects: structural and behavioral. On the structural side, an OODBS deals with implementation issues (e.g. storage). A structural system does not rigidly enforce encapsulation. Enforcing encapsulation proves to be a significant impediment to performance with regards to implementation. A behavioral view deals more with issues of semantics, maintaining integrity, and managing messages. A system often needs to support these two views simultaneously. One provides efficiency of storage, the other efficiency of object management. Incorporating both views adds to the incoherence of the object approach.

Given the disparity of approaches to implementing an object model, there is a strong rationale for establishing a measure of true object-orientation. Zdonik and Maier (1990) propose a 'threshold model' which establishes the following minimal features of an object database:

1. The provision of database functionality, meeting the minimal features outlined above.
2. The support of object identity.
3. Encapsulation, providing the basis for defining objects.
4. The support of composite objects.

Notably absent from this list is 'inheritance'. Also, some perceive inheritance as less fundamental than the above features, as it can be simulated with other language features. A weakness in the above definition, and one which permeates the literature, is ambiguity concerning the fundamental concepts, and a varying degree of adherence to a definition.

Although there is no clear consensus on the object model, there is agreement on the functions an OODBMS should provide. There are the minimal features that any DBMS should have, and there are those features which the niche applications will need. A group of distinguished researchers put together a set of thirteen Golden Rules as the minimum requirements to be satisfied by a database to be recognized as OODBMS (Atwood, 1991). These requirements are given below.

Requirements concerning object-orientation:

1. support for complex objects;
2. support of object identity;
3. support of encapsulation;
4. support of types or classes;
5. classes or types inherit from their ancestors;
6. no premature binding;
7. any computable function must be expressible in the language;
8. must be extensible.

DBMS requirements:

9. provide for persistence of data;
10. be able to manage very large databases;
11. accept concurrent users;
12. recover from hardware and software failures; and
13. provide for a simple way of querying data.

The group also identifies optional features that would make the OODBMS more attractive as:

1. multiple inheritance;
2. type checking and type inferencing;
3. distribution (distributed DB);
4. design transactions; and
5. versions (access to revisions of a single design).

Readers are also referred to another, somewhat similar, set of requirements proposed as the 'Third generation database system manifesto' by another group of researchers (Stonebraker *et al.*, 1990).

6.2.2 Taking on relational databases

Relational database management systems evolved in the 1970s. A relational model suggests the storing of all data in tables. The relationships among data items in different tables are expressed through references (by the inclusion of a foreign key) in other tables. Relational DBs were clearly an improvement over the cumbersome hierarchical and network structures employed by earlier databases. But there is a performance penalty, as complex objects must be assembled from primitive data structures every time the object is accessed (Stein, 1991).

While the relational model is currently most popular commercially, it clearly does not meet the needs of the niche markets. Certain features lacking (or supported weakly) in the relational model that could be more adequately provided by an OO approach are:

1. lack of correspondence with the real world;
2. inability to nest relations within tuples (the relational equivalent of a composite object);
3. lack of concurrency control and transaction management to deal with long transactions, such as those required by the CAD/CASE communities;
4. impedance mismatch between the Data Manipulation Language (DML) and application language – potentially rectified if querying is done with an OOPL;
5. lack of extensible data types for indeterminate data forms, e.g. as may be required for a multimedia database;
6. lack of schema evolution for capturing the temporal nature of the data and application.

Since engineering applications are becoming more complex in nature, a number of extensions have been proposed to extend relational databases. One such extension is that of an object implementation called a binary large object (BLOB). This is a data type that can hold image, sound or graphics in addition to character, numerical or logical data conventionally stored in RDMSs. A BLOB can be stored directly inside the database instead of in an external file, where it could be lost or altered. A presentation by Rowe and Stonebraker (1990) suggests that most of the features claimed by modern OODBMS are also implemented without sacrificing the advantages of the relational model.

Despite bold promises proposed by OODBMS, they must achieve a level of performance comparable to RDBMS in order to gain widespread acceptance and respectability – notwithstanding, of course, the lack of a coherent, well-defined object data model. Ideally, we would like performance measures which will clearly illustrate the differences between an object and relational system. Yet conventional DBMS performance measures – such as TP1 (Anon, 1985) – are not appropriate for an object model. Furthermore, the diverse target applications for OODBs can differ dramatically in their standards of performance, i.e. they make different demands upon a computing environment. Thus it is difficult to design performance benchmarks that will measure criteria relevant across the spectrum of possible applications. Therefore, Cattell (1988) suggests measuring operations on individual objects, and using an additive measure for composite objects (additive over the nested objects). These measures should be applicable to a relational model, to provide comparisons. Proposed benchmark operations are as follows:

1. *Lookup*: for example, retrieving an object given its name or finding the records with values within a specified range;
2. *Traversal*: finding all objects referenced by a given object, or (the inverse) finding all objects referencing a given object;
3. *Record Insert: including the update and time spent performing the above measures.*

Cattell's results using these measures were not impressive, however. His case is that OODBMS need a one-hundredfold improvement in speed to rival relational DBMS. In order to be justifiable, however, OODBMS do not necessarily have to equal relational models in performance. A reasonable speed would suffice in conjunction with added flexibility, ease of use, and features that meet the specialized computing needs of the niche markets.

In a more recent database benchmark experiment by Cattell and Skeen (1990), OODBs performed as well as relational DBs (better in some cases). A detailed discussion on comparison of relational and OO databases can be found in Loomis (1990).

6.3 Object identity

Object-oriented data models require an ability to make references to an object through an identifier. An identifier is a unique, system-wide reference which remains unchanged for the life of the object.

Identifiers should not be related to the physical location of the data, and should be implemented in such a way as to preserve the encapsulation of information. There are generally two dimensions to identifiers, representational (degree of support within the environment) and temporal (among transactions and structural reorganization). The ideal identifier provides complete value and structural data independence, and full locational independence. The most powerful identifier for the object model is considered to be attaching a surrogate value to each object, preferably one which reflects the above dimensions.

We would like to point out that the concept of identity is different from that of a primary key in a relation (in the relational model) (Maier and Zdonik, 1990, p. 9). A primary key is unique only within a relation, whereas object identity is unique across the database.

Relationships between objects can be indicated by using the object identifier as a reference in a related object. That is, if **aWorkstation** is related to **aWorker** object, the relationship can be established by storing the object identifier of **aWorker** in the object, **aWorkstation**. However, the association is activated by a method. A key, in the context of objects, is a property or set of properties that uniquely identify an object within a collection.

Relationships between objects do not have to be expressed through the identifier. A relationship is an expression of the abstract state of an object in regard to another object. Whether to model a relationship as an object or as a physical/logical link depends upon the application. Representing a relationship as a class may have certain advantages: for instance, in representing composite objects. An object's aggregates could be related via another object, called Contains, which implements the required semantics of that class's relationship with its aggregates.

6.4 Schema evolution

6.4.1 What is schema evolution?

The relational data model incorporates a static and global view of the world. The niche applications take a dynamic, temporal view of the world. Therefore, the data model must meet this representation requirement. The data model must be as flexible as the data itself, reflecting the real world dynamics. The OODB must support schema evolution. Under schema evolution, we will consider a model being developed at MCC (Microelectronics and Computer Technology Corporation) called Orion. Orion gives perhaps the most emphasis to schema evolution of the primary systems under development, and it utilizes a formalized framework. We will then consider other approaches to schema evolution.

Schema evolution is indispensable for CAD, OIS, and CASE applications. Changes develop frequently in the applications requiring data support, the usage of the database system, and the real world modeled by the database. These changes result from new patterns of database usage; specification changes in the application, often arising midway in the project lifespan; functional changes in the systems interfacing with the database; and finally, changes in the real world itself, requiring new data representation.

The notion of schema can be defined as a set of class definitions connected through an inheritance hierarchy, an *is-a* relationship. Two issues related to support of schema evolution are the incorporation of changes and propagation of changes. Incorporation includes application of changes at the class level. This entails verifying the integrity of the changes and making class modifications on a macro level. Propagation involves implementing the changes, i.e. imposing them on the instance, or micro level.

Some issues requiring consideration when implementing schema evolution are referential integrity, before and after propagation; automatic propagation versus manual; maintaining the integrity of concurrency and storage mechanisms; and propagation to composite objects. Potential problems are name conflicts, referential conflicts, and domain inconsistencies. Schema changes can occur on three levels:

1. *class definition*: adding, dropping, or modifying an instance variable or method;
2. *inheritance hierarchy structure*: that is, changing the inheritance among classes;
3. *class lattice*: such as adding or deleting a class, or changing the name of a class.

6.4.2 Orion

This model, as described by Kim and Chou (1988) and Banerjee *et al.* (1987), uses an axiomatic approach to managing schema changes. First, certain properties are defined which will be invariants of the class lattice; no changes may violate these invariants. Then, in the vein of mathematical theorems, definitions are established that guide the changes to the most meaningful way of producing them without violating the invariants. Keeping in line with a mathematical formulation, the rules and invariants are proved to be complete and consistent utilizing principles of graph theory.

The invariants are classified as:

1. *Class lattice*: no isolated classes, i.e. without super classes;
2. *Distinct name*: all instance and method names of a class are distinct;
3. *Distinct identity and origin*: all instance variables and methods of a class have a distinct identify and origin;
4. *Full inheritance*: all subclasses inherit all methods and instance variables from the superclass; and
5. *Domain compatibility*: all subclassed variables share the same domain as the superclass variable.

The rules, categorized below, may in certain cases be overridden by the user in conflict resolution.

1. *Default conflict resolution rules*: These select one inheritance option when there is a conflict of name or identity. They determine consistently what inheritance takes precedence when ambiguity or direct name conflict occurs.
2. *Property propagation rules*: These rules govern the changes to the properties of instance variables and methods. Properties pertain to such criteria as name, domain, default value, or coding. These rules in general propagate all changes

down the subclass chain, unless those changes incur conflicts in the subclass. In the case of conflict, the change is aborted at the subclass level.

3. *Directed acyclic graph manipulation rules*: These rules pertain to the addition and deletion of classes, as well as changes to the inheritance hierarchy. On the inheritance changes, the rules ensure that no classes are isolated, and by adding a superclass, that name conflicts will not require default name resolution. For class addition and deletion, the rules ensure that classes inherit from at least the default Object class (root class), and that system-defined classes remain intact.

4. *Composite Object Rules*: These regulate the operations performed on composite objects. A nested object may be taken out of the nested state; also an object may not become an instance of a non-composite class. Further, a dependent object may become 'disowned', i.e. a reference is kept intact but deletion of the owner will not cause deletion of the dependent.

6.4.3 Other approaches

Skarra and Zdonik (1986) approach schema evolution differently. Their scheme involves the changing of types *per se*, not just classes; a change to a type is treated similarly to a change in the class lattice. Types are defined by their operations (methods), properties (instance variables), and constraints (instance domain). The objective of their approach is to allow type change consistency, i.e. any type operation performed on the instances of a particular version to be well-defined for all subsequent versions. This consistency is supported through a version control mechanism, and by allowing for the inclusion of error handlers to type versions. These error handlers provide for situations in which an attempted action has been made invalid by a particular change. For example, if the domain of an instance variable changes and a method attempts to obtain a value outside the new domain, the error handler addresses this occurrence, providing either default values, or perhaps a message to the user or message-sending object. Errors are generally categorized as an attempt to access a property or operation which is undefined (resulting from the addition or deletion of properties or operations to a type), or a value which is unknown (resulting from changes in the domain of a property or operation).

Nguyen and Rieu (1989) discuss the propagation of changes among object instances. This is, perhaps, one issue that has not been adequately addressed in other OODBS. As mentioned above, propagation can be automated or it can be manual. Furthermore, it can be deferred or immediate. Deferred propagation is implemented in Orion, and is done only when an instance is accessed. It emphasizes performance, but requires the overhead of a permanent propagation mechanism. Immediate propagation (at the time the change is requested) emphasizes consistency and information preservation at the expense of performance.

The system of Nguyen and Rieu (1989) deals with the issues of incompleteness and inconsistency of changes in objects. They extend the notion of object to a partial definition of a relevant class. A relevant class represents a partial and meaningful design, characterized as a potential step toward a complete class definition. Regardless of the completeness or consistency, each instance is attached to exactly one relevant class. The relevant classes are not incorporated into a hierarchical or generalization – specialization relationship. They are implemented by the system to support incremental definition. Relevant classes are characterized automatically by the instance variables and constraints of a class in the lattice. The semantic rules governing a relevant class can depend on the data model and/or application dependent rules.

The propagation itself is performed using relevant classes and classification techniques (similar to that used in knowledge representation). First, schema modifications are performed, and their impact on relevant classes is assessed. The changes are defined and tested within the relevant classes. If the changes are correct, they are implemented on the appropriate set of instances. Since every instance belongs to one relevant class, updates are executed in sets, providing greater efficiency than object-at-a-time updates.

6.5 Versioning

The objective of extending functionality for multiple versions is to allow the user access to former versions based on arbitrary properties, working within a given context (e.g. developing a new version or maintenance on an old one). Beech and Mahbod (1988) outline the following requirements for version control for an object database:

1. *Creation*: Any object should be versionable, either implicitly or explicitly.
2. *Referencing*: Versions should be referenced either directly, e.g. through an identifier; or de-referenced through a generic reference according to some default criteria, which should be modifiable by the user.
3. *Causation*: Certain actions regarding versions should provide triggers to perform associated actions.
4. *Environment*: Versioning should be an integral part of a design environment, including supporting arbitrary queries.

The approach utilized in Orion is to have two types of versions, depending on the kind of operations that can be performed on them. The working version is a sort of master copy of the version. It is stable and cannot be updated. The transient version is a sort of experimental copy. It is derived from a working version and can be upgraded to a working version if it is deemed appropriate (e.g. after sufficient changes). Given the performance overhead of maintaining versions, classes are user-specified as versionable or not. A versionable class also has a generic object which serves as a data structure for holding information about each version. Thus each version has a generic object, containing system information, identifiers, next-version pointer, schema identifier, etc. A versioned object consists of a hierarchy of versions, called a version-derivation hierarchy.

A useful application of versioning not limited to design applications is schema versioning. There are multiple approaches to accommodating this. The versions of the schema approach are to view the entire class lattice as a versioned object. The views, of the schema approach are to provide dynamic, user-defined views, rather than versions, of the schema. The versions of the class approach are to view each class as a versioned object.

Kim and Chou (1988) utilize the versions of schema approach. Their model consists of seven rules, which in part extend the Orion model for versioning objects. Their model constructs a hierarchy of schema derivations, which the rules govern. A schema version is either working or transient, as described above. The rules determine what objects a particular version has access to in relation to its position in the inheritance hierarchy. For example, a 'child' schema version cannot update its parent version; a schema version cannot be deleted if it has 'children' versions; and a version can update all of its objects, including those inherited, but the updates are only visible (accessible) to that version and its descendants. One difficulty with this approach is in the system overhead to support object updates inherited from parent versions.

6.6 Storage

In general, most OODBS are organized into a storage manager (SM) and an interpreter. The interpreter implements the SM for physical disk access and provides the operational semantics of the data model, executes methods and enforces encapsulation. The SM is concerned with the organization of objects in secondary storage, data movement between disk and main memory, recovery, concurrency control, object creation, updating, etc. Another approach for providing object persistence is to have a layer of recoverable virtual memory, wherein object semantics are described in the layers above the virtual memory. A key question in implementing storage optimization methods is the tradeoff between keeping strict encapsulation to preserve true object orientation, or to sacrifice some encapsulation for the sake of performance, where performance is measured by I/O time.

While there is concern over the lack of performance via query optimization in OODBS, there is a case for storage efficiency having a greater impact on performance than query optimization. The niche applications, particularly engineering, are generally not query-intensive. These applications are more prone to rely on fewer queries, with more time spent in the output of each. Under this scenario, the issue of object traversal has a higher impact on performance than query optimization. Object traversal is optimized through more efficient disk storage, e.g. reducing page faults and improving object clustering. Thus the storage mechanism incurs greater weight on system performance.

6.6.1 Considerations for storage design

Some important questions related to storage design are discussed below.

How much should the storage manager know about the semantics of the data model?

Weak semantics in the SM enables simplified implementation, greater generality and reusability for multiple data models, and centralized semantic functions (in the interpreter). Yet the tradeoff is that the SM must pass the buck on low-level maintenance issues such as garbage collection, index maintenance, and constraint enforcement. With more semantics in the SM, potentially more storage optimization and more efficient enforcement of constraints can be achieved, as can the ability to maintain auxiliary access paths for query optimization.

Should a custom storage manager be built, or an existing one used?

Clearly, it would be more efficient to use an existing SM, such as one used for a relational database system. However, it is questionable whether an existing system can be fine-tuned to match the requirements of an OODB. Furthermore, this approach may best be used in the cases of extended relational systems. Otherwise, for efficiency purposes, a customized manager would provide more control.

How should an object's state be represented in memory?

This question gets more complicated when the SM knows something about the structure of the objects. The straightforward approach is to have the state of the object represented contiguously, with component objects' states separate. Otherwise, the object state can be subdivided and grouped according to their particular type; thus all fields of a certain type will be stored together. This lowers search-time when retrieving all instances of a certain type, but retrieving all fields of a certain object is costly. A combined approach is possible, but this presents additional overhead on updating.

A system that supports large objects (those whose states will generally occupy more than a page of memory), would not want to represent states contiguously. Contiguous storage of large objects could lead to excessive fragmentation of memory and possibly to time-consuming data compaction. Rather, a tree structure is used, with indexing on byte (or array) position, to break the state into segments, allowing for partial retrieval of a state.

How should objects be clustered on disk?

Clustering objects closely on disk that are likely to be used together or refer to each other, decreases access time when reading into memory. However, since objects can (and usually do) have multiple references, optimal clustering is a complex task. A consideration is how much clustering should be automatic, and how much manual. Approaches include the gathering of statistics on traversals between objects to evaluate which objects should be stored together; also the allowing of the user to specify whether objects be clustered as breadth- or depth-first. Another approach is to provide a unit of clustering, called a segment, into which objects are placed manually and/or automatically. When part of the segment is needed, the whole segment is loaded. These segments can form a flat partitioning of the object space, or segments can be nested within each other.

How should object identifiers be implemented?

One technique involves using either physical or logical addresses, which are de-referenced through an object table. The table facilitates easier object movement, since the identifier (required to be unique system-wide) is not tied to a physical location. Yet this method introduces another level of indirection, hence more time de-referencing. Further, the table requires heavy usage for concurrency control and recovery, creating a potential I/O bottleneck. A compromise involves using an entity identifier, mapped through an entity directory, wherein the identifier may incorporate a physical address, providing a hint as to where the object is stored. The state of the object always contains the identifier, perhaps in the form of a surrogate.

Another approach is to maintain a hash table of object identifiers to memory addresses. While it may be more time-consuming than direct memory access, it is more cost-effective than de-referencing through an object table. Similar to a hash table is a technique called 'swizzling'. This technique takes the disk-based address and swizzles it into either a short identifier for objects in memory, or to main memory addresses. This has the advantage of mapping from the object reference to the address only once, similar to indirect addressing. Some complications involve when to do the translation so as not to waste time on unused references; this can be circumvented by swizzling the references when they are first accessed within the main memory.

6.6.2 Persistent objects

Programming languages stress transient data with rich data types, while databases work with persistent data and limited data types. This has led to the two-pronged approach of implementing a storage manager and an interpreter to manage objects.

Persistent data is data that is required to exist outside the scope of any particular program run. In general, persistence can be thought of as a requirement of data wherein the data will become inaccessible only at a time when some predetermined criteria have been met. The existence of data and its integrity should not be vulnerable to hardware or software problems. There are degrees of persistence, ranging from the transience concept to data that outlives the program. Generally, DBMS are concerned with the degrees of persistence wherein the data will remain intact between executions of the program.

A key consideration in managing persistent objects is how to name, or identify, objects. As discussed in section 6.3, a unique, system-wide identifier is required. As the potential requirements for object access are enormous, a two-level naming scheme may be appropriate, as persistent objects should require names longer than transient objects. Persistence must apply to all objects involved in all composite objects, to prevent dangling references in the persistent storage. Also, persistent objects should resemble transient objects when loaded into memory.

Conceptually, the persistent memory is a collection of containers for the object state. The object state can be thought of as consisting of distinct components such as: the identifier of the object's class; an instance identifier; the timestamp of the last write to the object; an indicator of whether a type is primitive or composite; and a listing of the instance variables.

A fundamental question is: what makes an object persistent? One way to answer this is in specifying when an object becomes persistent. Possible approaches are: at creation time; upon instruction by a message; by having a persistent class and transient class; or when an object is stored in persistent storage (Zdonik and Maier, 1990). The approach used will largely depend on the system's design and how it relates transient objects with persistent ones.

Immutable objects

Immutable objects refer to those whose state cannot change, as no operations 'mutate', that state. These objects are created in a certain state and remain that way for their lifespan. This concept can be relaxed so that an object can acquire immutability only after a certain time, before which it is mutable. Thus this type of object has mutating operations which become invalid after a certain point, at which point its state is fixed. At the point of becoming immutable, it becomes read-only.

Immutable objects lend efficiency to a shared, persistent object base. They do not participate in concurrency control, and, once read into local memory, can be read at minimal cost. Immutability also enforces integrity constraints on a sharable object base. In principle, the concept of immutability establishes a distinction between the container for the value and the value itself. This is an important notion in persistent data sharing, as sometimes it is desirable to share a value, and other times to share a container. For these reasons, Low (1988) suggests making as many objects as possible immutable.

6.7 Querying

Query language in a DBMS provides the syntax for all possible types of access (to a database) generated by the underlying data model. Since the OO data model is richer than the relational data model because of additional constructs such as methods, class inheritance, arbitrary data types and set-valued attributes etc., the query language is bound to be more complex than that for the relational model.

Query capabilities are considered one of the weaker aspects in operational OODBs. With the standard set by the efficient relational query tools, the object approach has a lot to contend with. Most current OODBs do not account for the fundamental object concepts in querying. Others provide a new language which is backward-compatible with SQL, or simply provide an SQL interface.

A query language for OODB has two main obstacles (Cluet *et al.*, 1990). The first is the principle of encapsulation. Encapsulation hides object structure, thereby making it difficult to determine how a message is being implemented. This property affects query optimization negatively. An approach to getting around this limitation is to allow the query optimizer access to the object's structure, to help evaluate alternatives. This approach compromises the encapsulation concept, which may conceivably be tolerated depending on the performance improvement obtained through optimization. Some systems interpret the optimizer as a 'trusted' component of the storage manager, wherein it has built-in features for enforcing this trust. Nevertheless, dealing with the broad range of possible data structures and algorithms, obtaining efficiency from extensible data types could easily cost more than it saves.

The second obstacle is the richness of object-oriented structures, including extensible data types and composite objects, which causes difficulty in reconciling complex data structures with the simplicity expected from a query language. In dealing with extensible data types, new types create operators which have indeterminate algebraic properties. Composite object querying has been treated largely in the same manner as other non-first normal form models. However, there has not been much success in implementing an efficient query language for these forms.

Yet another difficulty is with the built-in relations among objects, implicitly defined by the given set of methods. These pre-defined links limit the number of access paths which a query can choose from for optimization. This problem can be alleviated with the use of indexes. Also, support for schema evolution could provide user-defined methods which enable the user to create links more appropriate for a particular application.

While there are algebras for expressing queries over sets of objects, in addition to the problems mentioned, the main problem is query optimization. Many attempts at optimization methods have focused on extending principles developed for the relational model and incorporating them into algebras created for objects. As this approach may produce satisfactory results, it may not make full use of the object-oriented concepts and may even violate them. The situation is further complicated by the fact that many existing DMLs for OODBs return results that do not have the same structural properties as the original database, i.e. they do not support the closure property (Alashqur *et al.*, 1989). This causes problems when trying to nest queries. Furthermore, query optimization strategies will have to consider class storage implementations. Some classes will be stored as a B-tree, and others through a hash table. User-defined operations and abstract data types, introducing arbitrary complexity, may have to be ignored by an optimization strategy.

As mentioned earlier, some target applications for OODBs do not require extensive querying, and hence optimization is not critical to performance in these cases. However, this certainly does not imply that some database systems will not require a combination of capabilities, serving perhaps both business and design applications. For these purposes, an OODBMS would ideally interface smoothly with existing database systems.

6.8 Commercial OODBMS

Many vendors have released OODBMS in the past few years. Some of them are listed below:

Company	Product name
Objectivity, Menlo Park, CA	Objectivity/DB
Object Design, Burlington, MA	Object Store
Ontologic, Burlington, MA	ONTOS
Servio Corporation, Alameda, CA	Gemstone
Versant Object Technology, Menlo Park, CA	Versant
Altair, Le Chesnay Cedex, France	O2
Itasca Systems, Minneapolis, MN	Itasca

It should be noted that this is not an exhaustive list of vendors; there are others. Readers are referred to the directory compiled by Salmons and Babitsky (1991) for a more comprehensive list.

The commercial products are quite sophisticated with some marketed as distributed OODBMS. Many of them offer add-on products such as bridges to relational databases, visual programming tools, interfaces to expert systems, support for multiple languages such as C++, Smalltalk, C, COBOL etc., and a variety of development tools. Most of them are also available across various hardware platforms.

6.9 Conclusion

In this chapter, we have discussed the status of OODBs regarding select criteria most relevant to its rivalry with the relational model. Clearly the technology has many unresolved problems and different approaches to each problem. Many of these problems result from the lack of a coherent, formalized framework or data model. This in turn is reflective of the fundamental inconsistency of a database in an object context. Nevertheless, in the long run, OODBs have strong potential for acceptance and respectability, with the key ingredient being a formalized framework.

Schema evolution is a requirement of the niche markets which OODBs seem most suited for. It reflects applications in which the data has a temporal quality. Support for schema evolution enables the database to change consistently with the application. This enforces a closer correspondence between the system and the real world – a primary advantage of the object-oriented approach. The Orion system at MCC currently emphasizes this need more than other systems, and it utilizes a formalized approach, thereby lending credibility. Related to the schema evolution is versioning. This too reflects a closer correspondence with the real world and is facilitated through the object approach.

Efficiency is clearly a two-pronged issue. It hinges on effective storage management as well as query optimization. However, querying will most likely have to wait for the object model before optimization is guaranteed. Storage mechanisms are more mature, with relational storage methods more applicable. Acceptance will also be affected by the degree to which an OODBs will meet user requirements and the degree to which the system will integrate smoothly with existing systems, particularly those it will interact with.

Further potential exists from the continued integration of diverse principles in disparate research directions. In particular, on the integration of knowledge bases and databases, we see OODBs as providing a convenient and effective structure, pending further refinement.

References

Alashqur, A. M., Su, S. Y. W. and Lam, H. (1989) OQL: A query language for manipulating object-oriented databases. In *Proceedings of 15th International Conference on Very Large Databases* (eds. G. Wiederhold and P. Apers), Morgan Kaufman Publishers, San Mateo, CA.

Anon (1985) A measure of transaction processing power. *Datamation*, **31**(7).

Atwood, T. M. (1990) The object-oriented database system manifesto: A consensus from academia. *Hotline on object-oriented technology*, **1**(3), 6–9.

Atwood, T. M. (1991) The case for object-oriented databases. *IEEE Spectrum*, April, 44–47.

Banerjee, J., Kim, W., Kim, H. J. and Korth, H. F. (1987) Semantics and implementation of schema evolution in object- oriented databases. In *Proceedings of Sixth ACM SIGACT-SIGMOD-SIGART Symposium on Principles of Database Systems* (eds. I. Traiger and U. Dayal), ACM Press, Baltimore, MD.

Beech, D. and Mahbod, B. (1988) Generalized version control in an object-oriented database. In *Proceedings of Fourth IEEE Conference on Data Engineering*, IEEE Computer Society Press, Los Angeles, CA.

Cattell, R. G. G. (1988) Object-oriented DBMS performance measurement. *Advances in Object-Oriented Database Systems: Second International Workshop on Object-Oriented Database Systems* (ed. K. R. Dittrich), Springer-Verlag, New York.

Cattell, R. G. G., and Skeen, J. (1990) Engineering database benchmark. Technical Report, Database Engineering Group, Sun Microsystems, Mountain View, CA.

Cluet, S., Delobel, C., Lecluse, C. and Richard, P. (1990) RELOOP, an algebra-based query language for an object-oriented database system. *Data and Knowledge Engineering*, **5**(4).

Date, C. J. (1990) Introduction to Database Systems, 5th edn., Addison-Wesley, Reading, MA.

Gupta, R., Cheng, W. H., Hardonag, I. and Breurer, M. A. (1989) An object-oriented VLSI CAD framework. *IEEE Computer*, May, 28–37.

Gupta, R. and Horowitz, E. (1990) *Object-oriented databases with applications to CASE, networks, and VLSI CAD*. Prentice Hall.

Kim, W. and Chou, H. (1988) Versions of schema for object- oriented databases. In *Proceedings of 14th International Conference on Very Large Databases*, Morgan Kaufman, Los Angeles, CA.

Kim, W. and Lochovsky, V. (eds.) (1989) *Object-oriented concepts, databases and applications*. Addison-Wesley, Reading, MA.

Loomis, M. E. S. (1990) The basics; ODBMS v. relational, etc. A series of articles, *Journal of Object-Oriented Programming*, **3**(1), 77–81.

McLeod, D. (1988) A learning-based approach to meta-data evolution in an object-oriented database. *Advances in Object-Oriented Database Systems: Second International Workshop on Object-Oriented Database Systems*, (ed. K. R. Dittirch), Springer-Verlag, New York.

Nguyen, G. T. and Rieu, D. (1989) Schema evolution in object- oriented database systems. *Data and Knowledge Engineering*, **4**(1).

Rowe, L. A. and Stonebraker, M. R. (1990) The POSTGRES data model. *Research Foundations in Object-Oriented and Semantic Database Systems*, (eds. A. F. Cardenas and D. McLeod) 91–110.

Sadri, F. (1989) Object-oriented database systems. In *Proceedings of 13th IEEE International Computer Software and Applications Conference*, IEEE Computer Society Press, Los Angeles, CA.

Salmons, J. F. and Babitsky, T. T. (1991) *1991 international OOP directory*. COOT Inc., NY.

Skarra, A. H. and Zdonik, S. (1986) The management of changing types in an object-oriented database. In *Proceedings of OOPSLA '86* (ed. N. J. Meyrowitz), ACM Press, New York.

Stein, J. (1991) OO DBMS: Object-oriented technologies for complex data-management systems. *SunWorld*, May, 73–84.

Stonebraker, M. (1991) Object management in postgres using procedures. *On Object-Oriented Database Systems* (eds. K. R. Dittrich, U. Dayal and A. P. Buchmann). Springer-Verlag, Berlin, 53–64.

Stonebraker, M. *et al.* (1990) Third generation data base system manifesto. In *Proceedings of 1990 ACM-SIGMOD Conference*.

Zdonik, S. (1989) Directions in object-oriented databases. In *Proceedings of 13th IEEE International COMPSAC*, IEEE Computer Society Press, New York.

Zdonik, S. and Maier, D. (eds.) (1990) *Readings in object-oriented database systems*, Morgan Kaufmann, San Mateo, CA.

7 Comparing object-oriented programming languages

M. GADRE

7.1 Introduction

While the concepts behind object-oriented programming (OOP) have been around for a few years, it is only recently that this software engineering technique has become widely accepted. In addition, with the rapid rise in hardware performance, the traditional drawbacks of object-oriented programming systems (OOPS) (such as larger memory requirements and slower speed) have diminished in significance.

In response to the increased need for object-oriented programming languages and environments (and also driving this need) a variety of such languages have been introduced. In a few short years the number of choices available to the developer of object-oriented systems has greatly increased. Saunders (1989) provides a high-level overview. Stefik and Bobrow (1985) examine object-oriented programming languages from the perspective of artifical intelligence (AI).

This chapter is a summary of the features of major object-oriented programming languages available today. While syntactical and other differences exist, most of these languages are conceptually quite similar. A majority of the languages are extensions to either LISP or C. Because of its greater availability, efficiency as well as ability to run on most platforms without requiring special hardware, C and its object-oriented extensions are better suited for developing manufacturing related applications. Hence this chapter will present a more detailed study and comparison of two popular extensions to the C programming language: C++ and Objective-C. The discussion will illustrate how these two languages provide for the major features of object-oriented programming. In conclusion, some guidelines on selecting a language for particular applications are discussed.

7.2 Smalltalk

The first complete implementation of an object-oriented programming language (OOPL) was Smalltalk, and it is considered to be the purest implementation of the object-oriented philosophy. Smalltalk was one of the primary forces behind the dissemination and popularization of object-oriented software concepts. This influence can be seen in many of the object-oriented extensions to conventional languages that exist today. However we will restrict ourselves to only a very brief overview here since a detailed analysis is beyond the scope of this chapter. Interested readers can refer to other relevant publications for more information, such as Goldberg and Robson (1983); Digitalk (1986) or Rettig (1987).

Smalltalk is an object-oriented programming *environment*, not just a language. It provides the user with powerful features such as browsers, bitmap editors and inspectors (Rettig *et al.*, 1989). In Smalltalk everything, including fundamental elements such as integers and characters, are objects. As a result, doing even elementary operations such as adding two integers involves passing a message from one integer object to another. This high degree of consistency makes it difficult for the Smalltalk programmer to 'cheat' on the implementation of OOP concepts. While this could initially be frustrating for the novice object-oriented programmer, it soon forces the developer into applying and benefitting from the object-oriented paradigm.

This purity comes at a price, however. Smalltalk is resource-hungry and runs more slowly than other 'less complete' object-oriented environments. The Smalltalk environment is delivered with a large set of class libraries. As a result, one of the primary hurdles to object-oriented programming, the necessity of having mature class libraries, is eliminated. Hence the Smalltalk programmer can quickly begin developing non-trivial applications.

7.3 Object-oriented extensions to LISP

Of all the languages that have been extended to provide object-oriented features, LISP is perhaps the most popular. Not only has the object-oriented paradigm been successfully incorporated into LISP, but there are some clear innovations.

7.3.1 Innovations

A brief description of the major innovations introduced by the LISP-based OOPLs follows. For further details the reader is referred to Tello (1989) and Rettig *et al.* (1989).

Mixins

Mixin is the LISP version of multiple inheritance. It allows a class to inherit from more than a single other class. The inheritance can be in a tangled hierarchy or a network. For example, suppose a certain set of classes have some common characteristics. These can be abstracted into a single class. This class can then be 'mixed in' with the others.

Method combinations

Method combinations first appeared commercially in the Symbolics Flavors system. It allows the user to define the manner in which methods from different components are to be combined in order to produce the new behavior. The default, of course, is to ignore all but the most recent implementation of the method. (This is, for example, the case with Objective-C where a method can be overwritten in subclasses and only the method appropriate to the class (or subclass) will be invoked.)

With the Flavors system however, the *user* can specify the manner in which the methods from different components are combined, invoked and/or ignored. This is done via the specification of '*primary*' and '*daemon*' methods. The primary method controls the major functionality of the method while the daemon methods perform subsidiary tasks. Daemon methods are of two types – 'before' and 'after' methods. This terminology is based on the order in which the methods are executed. First, all the before methods are called, then the primary method and then all the after methods.

It is important to note that each component method is passed the same arguments as were passed the combined method. However the return value from the combined method is the same as that from the primary method. All the return values from the daemon methods are discarded. While this may seem like needless sophistication to some, the fact to be appreciated is that this provides a very high degree of reusability. The user has no reason to recode functionality that already exists. He or she simply has to find the set of component methods that provide the closest fit to the requirements at hand and then mix them in the appropriate order.

Multimethods

Multimethods are functions that can be messages to a variable number of objects. In the case of an Objective-C message expression, for example:

  [window display];

the message is sent to a single receiver and the class to which it belongs is responsible for its implementation. In the case of multimethods, however, the receiver is also just another argument, of which there could be many. Each argument could be of a different class or type. The order in which the objects appear will determine the method combination desired.

7.3.2 Example dialects

Some of the dialects of object-oriented LISP are briefly discussed next. Micallef (1988) performs a more detailed analysis of how these languages support different object-oriented programming requirements.

Flavors

This is the precursor of the LISP-based object-oriented programming languages. It was developed at the Massachusetts Institute of Technology by the LISP Machine group in 1979. Classes are called '*flavors*' and are defined using the keywords **defflavor, defmethod** and **make-instance**. While the detailed nuances of the syntax are not discussed here, the manner in which daemon methods are defined is noteworthy. Suppose we have a class called **Window**.

```
(defflavor Window
  (name xLoc yLoc ptrLoc)
  ( ))
(defmethod
  (Window :before :take_user_input)
  ( )
  (clear_display))
(defmethod
  (Window :take_user_input)
  ( )
  (process_user_input))
(defmethod
  (Window :after :take_user_input)
  ( )
  (unmap))
```

Sending the message

```
(send 'Window :take_user_input)
```

would result in the following sequence of execution:

```
clear_display
process_user_input
unmap
```

ObjectLISP

This dialect has some interesting features that are useful for developing OOPS. In the ObjectLISP system the relation of a class to its instances is the same as that of the class to its subclasses. This simplifies the syntax in that the nesting of closures is used for both instantiating an instance as well as specializing the class. One of the byproducts of this approach is that during prototyping the developer can use a class as a prototype instance, or an instance as a prototype class.

ObjectLISP does not have any special syntax for messaging. Messages are sent using essentially the same syntax as that for calling any CommonLISP function. The user can create and modify objects on the fly when programs are running. In addition, inheritance operates dynamically. This means that changes to the state of a superclass will percolate downwards at the time of the change (Tello, 1989).

XLISP

XLISP was conceived as a tool for introducing object-oriented programming and related concepts. It is an extremely simple system that is available on a wide range of platforms (Rettig *et al.*, 1989). There are only two built in classes – **Class** and **Object**. New classes are defined by sending the new message to **Class**. This can be followed by an optional list of class and instance variables. Only single inheritance is supported.

```
(setq Window
  (send Class
    :new' (name xLoc yLoc)))
```

Methods are defined on this new class by sending it the :**answer** message. This includes the message selector, argument variables as well as the LISP code that is the implementation of the method.

```
(send Window :answer :greeting
  '(princ "Hello World"))
```

Instances are created by sending the :**new** message (with optional arguments) to a previously defined class. (The classes are defined by sending :**new to Class**.) XLISP first creates the new instance and then sends it the message :**isnew** with the arguments supplied to :**new**. This means that the developer can place initialization code in the **isnew** method.

LOOPS

LOOPS was developed at Xerox PARC and released in 1983. While Common-LOOPS was simply an object-oriented extension to CommonLISP, LOOPS is an object-oriented, high-level AI language with many facilities for developing advanced AI applications (Tello, 1989). LOOPS was designed to support a multiple paradigm framework. It currently supports the procedural, access-oriented, object-oriented and rule-based paradigms; see also Bobrow and Stefik (1983).

The syntax is a little different from the other LISP-based OOPLs discussed above. User defined classes are referenced using the $ symbol as follows:

```
($ GUIElement)
```

The syntax for sending a message called **create** to a user-defined class called **Window** will be as follows:

```
(← ($ Window) create 'myWindow)
```

The new instance will be called **myWindow**. Methods in LOOPS are created using the **DefineMethod** function (Tello, 1989).

(DefineMethod class selector argsOrFunc expr file methodType)

The method implementations are as follows:

(Method ((ClassName selector) self arg1...argN) ... body)

The **self** variable points to the class to which the message will be sent. LOOPS provides multiple inheritance as well as facilities for managing rules. Rules are organized into rulesets which have different evaluative control structures. Based on the programming paradigm, these rulesets can be invoked in different ways. In the object-oriented paradigm this is done by sending a message to the object containing the ruleset.

7.4 Object-oriented extensions to C

In order to make some of the benefits of OOP available to applications programmers without incurring the large overheads associated with environments such as Smalltalk, 'hybrid' languages have been developed. These languages allow the programmer to write code in the conventional, base language. This has the benefit of optimizing performance where the rigid adherence to OOP principles is a handicap. An obvious example is performing computation intensive operations. In such a situation, the overhead of messaging for basic arithmetic operations is a liability. Hybrid languages support encapsulation and (varying degrees of) dynamic binding, but allow the programmer to 'cheat' and deviate from (and perhaps lose the benefits of) object-orientation. With such languages, the responsibility of adhering to these principles is on the software engineers and managers. The environment may not guarantee adherence (as it will in the case of Smalltalk).

A language that lends itself easily to such extensions is C. After LISP, C has perhaps spawned the most number of object-oriented dialects. Even in conventional C it is possible to quite easily apply many of the software engineering principles of OOP, such as code localization, modularity and data hiding. Good software engineers have been applying these principles either consciously or unconsciously for some time. Examples of C constructs that support such principles are **static, struct, typedef** and so on. Interested readers are referred to Brumbaugh (1990) for examples. In addition, the facilities provided in the language for dynamic memory allocation and manipulation make it possible to apply the dynamic aspects of OOP.

As a result, the compilers provided for such object-oriented extensions to C are merely preprocessors that convert the extended syntax to C, while ignoring the statements written by the programmer in conventional C. The resulting file is then passed to the C compiler available on the system to produce the object modules which are then linked to produce executables. (This is changing now with systems such as Turbo C++ and Zortech C++ which are native C++ compilers and provide C++ source debuggers.)

C++ and Objective-C are two popular hybrid languages with C++ being the more successful commercially. Each provides varying degrees of object-orientation that stem from the fundamental philosophies driving the languages. Each stems from different underlying philosophies – C++ redefines the C language itself while Objective-C adds object-oriented capabilities to an unchanged base C language.

A key decision in both cases has been to make both languages compatible with C. Both Objective-C and C++ are supersets of the C language. This preserves the usefulness of many million lines of mature C code. It also allows the programmer access to the power, efficiency and flexibility of C. It is possible in both languages to use C to optimize the performance of machine level programming while using the object-oriented extensions to encapsulate such behavior in methods and objects. The extensive sets of C libraries and tools can still be used, providing developers the option of moving to a far more powerful paradigm without giving up years of experience and development.

7.5 C++

C++ has been designed and implemented by Bjarne Stroustrup, a member of the Computer Science Research Center at AT&T Bell Laboratories in Murray Hill, New Jersey. It was developed to support event-driven simulations for which Simula67 was too inefficient. It was first installed outside Stroustrup's language development group in July, 1983 (Wiener and Pinson, 1988) and is yet to be standardized.

C++ is a redefinition and extension of the C language. The name 'C++' signifies the evolutionary nature of the language – '++' is the increment operator in C (Stroustrup, 1987). Apart from the obvious dependence on C, the C++ class concept was inspired by Simula67 (with derived classes and virtual functions). In addition, the BCPL comment convention '//' has been introduced (Stroustrup, 1987). While C++ owes much to C, some aspects of C++ have been accepted into ANSI C: for example, the '**void** *' pointer type and function prototypes.

According to Stroustrup (1987), the difference between C and C++ is primarily in the degree of emphasis on types and structure. C++ is more expressive than C, but in order to gain that benefit, the developer must be more attentive to the types of objects.

It is important to note that C++ is not simply the addition of features to the base C language. It is a natural and consistent extension to C. The changes include basic enhancements to the operators and syntax of C, as well as facilities for object-oriented programming. Let us look at some examples.

7.5.1 Examples of how C++ enhances C

The scope resolution operator

The scope resolution operator '::' allows access to externally declared variables. The major use and benefit of this operator is with classes. In the case of standard

C code, it is possible to use hidden global names using the :: operator. The following code fragment will be instructive:

```
filename: xmpl.c
    int x;   /* file scope: variable global to all functions in xmpl.c */
    int func( )
    {
        int x;           /* scope within func( ) */
        :: x = 10;       /* set the global variable */
        x = :: x;        /* assign value of global to local variable */
        . . .
        return x;
    }
```

It is quite possible to misuse the scope resolution operator. See Stroustrup (1987, p. 42) for examples.

Avoiding the preprocessor

The use of macros (#**define**) can be reduced in C++. Manifest constants can be defined using **const** or **enum**. Use **inline** in order to avoid function call overhead. The **const** specifier applied to a basic type yields a type whose value *cannot* be changed after initialization, which *must* be done at the point of declaration. For example:

```
    const int delta = 1;
```

The position of the **const** modifier when declaring pointers is important. There are two possibilities:

```
    const float * ptr;
    float * const ptr;
```

In the first case, the pointer value is modifiable but the value being pointed to is not. In the second case, the pointer is unmodifiable but the value pointed to can be changed.

Since the C preprocessor does not understand C syntax, using it to define macros requires extra parenthesization and prevents adequate type checking. The adjective **inline** is an indication to the compiler that the function be compiled as a macro. The declaration:

```
    inline int max(int x, int y) { return ((x > y) ? x : y); }
```

provides type checking and is far more robust than:

```
    #define MAX(x, y) (x > y ? x : y)
```

which will collapse in a situation such as **MAX**(a + b, c + d). An **inline** function is expanded by the compiler wherever it is called. The drawback to the **inline** specifier is that the compiler could choose to ignore it if it decides that the function is 'too complex'. The rules for determining the degree of complexity are not clear. (It might help to compare the behavior of **inline** with that of **register** in C; both are requests to the compiler, not directives.) In general, **inline** should be used only for short, commonly used functions, where function call overhead would be significant.

Function declarations and implementations

Function prototypes (now a part of ANSI C) enhance type checking and allow the implementation details to be hidden from the interface. In addition to this, C++ provides the following enhancements to the manner in which programmers can declare and implement functions.

Overloading function names:

This allows the same name to be used for functions with different types and numbers of arguments. This is useful in cases where we have functions performing the same task on elements of different types. It is also possible for the return types to be different. For example, the function **strcat** ( ) – from the standard library **string.h** – accepts two arguments of type (**char ***). If one wanted to use it for arguments of some other user-defined types then casting of the variables or some other manipulation is required. An example of where such a situation could occur is in graphical user interfaces (GUI). A text string that is displayed on the screen could contain information on the font used, its size and position in addition to the ASCII character string. By simply defining the different function prototype, the function can be overloaded. Suppose we had a user defined type called '**guiStr**'.

```
char *strcat(char *, char *);      // from string.h
guiStr strcat(guiStr, guiStr);     // user defined
```

Default values:

Default values can be assigned to the trailing set of arguments in a C++ function. This allows the programmer to invoke the function with only a subset of the total set of parameters. Default values are used for the unspecified parameters. For example, a function declared as

```
void func(char *filename, char *option = "read");
```

can be invoked in either of two ways:

```
func("xmpl.c"); // 2nd parameter defaults to "read"
func("xmpl.c", "delete");
```

Apart from the reduction in the possibility of the errors involved in having to invoke the function with more than required parameters, this facility can have some profound benefits. Suppose that there is a function:

```
importantFunc(int x);
```

that performs an important task and is widely used in an application. It is decided that it needs to be enhanced and will require more parameters. However, since it is used so widely, it will continue to support the older behavior. The programmer can ensure that the other users will not be affected by giving default values to the added parameters. The enhanced implementation will use these default values to perform the task of the older definition:

```
importantFunc(int x, int y = 0.0, char *behave = "old_way");
```

Functions with an unspecified number of arguments:
This is required when it is impossible to specify the number and type of all the parameters expected in a call. The ellipsis, '. . .', is used to specify an unknown number and type of arguments. For example:

```
func(. . .);
```

This suppresses type checking and promotes flexibility in the function interface. However, unless the function is implemented with extensive error checking and recovery mechanisms, it could be a recipe for disaster.

Call-by-reference:
A variable can be called by reference by using **type&**. This allows the function to directly manipulate the variable without requiring the manual navigation of pointers. A *reference* is an alternative name for an object. For example:

```
void func(int x, float &y)
{
  x++;          // increment only local copy
  y *= 2.0;     // change actual argument
}
void anotherFunc( )
{
  int a = 1;
  float b = 3.0;
  func(a, b);   // a = 1, b = 6.0
}
```

While the same effect could be achieved by the passing of the pointer to b and its subsequent manipulation, the above is easier to use and read. Programmers with experience in a language like FORTRAN will appreciate the more readable style.

Memory management

Instead of using **malloc( )** and **free( )**, C++ provides the **new** and **delete** functions. These functions have the ability to determine the type of memory space requested and return a pointer of the appropriate type. For example, instead of int *x = (int *) malloc(sizeof(int) * 10);

we can say

int *x = new int[10];

which is much simpler to use and read. The free store allocated by **new** must by returned by explicitly using **delete**, there is no 'automatic garbage collection'. **delete** can be applied only to a pointer returned by **new**. There are additional error recovery mechanisms available. When **new** fails, it calls the function pointed to by _new_handler, which can be directly set or the **set_new_hand-ler( )** function can be used to set it.

The above set does not include all the enhancements to C provided by C++. The examples have been selected mainly to give the reader a flavor for and an appreciation of the new features. A complete list along with all the variations and pitfalls will require a complete text book or reference manual and the interested reader is referred to one (Stroustrup, 1987; Wiener & Pinson, 1988; Pohl, 1989; Ellis and Stroustrup, 1990).

7.5.2 Facilities for object-oriented programming

Of fundamental importance to the development of object-oriented software in C++ is the **class** construct. This is a superset of the definition of the **struct** construct in C. In C++ the **struct** is simply a special case of **class**. As mentioned above, it has been possible to apply many aspects of object-oriented programming in traditional C. The **struct** has been a way of achieving data encapsulation. The **class** in C++ enhances that feature to allow encapsulation, as well as control the visibility and accessibility of the data *and* the procedures. The visibility can be controlled through the use of the keywords **private, protected** and **public**. When an object is declared, C++ automatically calls the '*constructor*'. This is a method that has the same name as the class with any number of arguments. When an object goes out of scope or is explicitly deleted using the **delete** function, the '*destructor*' is called. This also has the same name as the class but is preceded by a '~'. Member functions are declared in the same manner as in C, with the return type followed by the function name.

Before we discuss examples, let us briefly review some terminology that is unique to C++. Classes are also referred to as *abstract data types* (ADTs) and are considered to be user-defined extensions to the existing data types available in the language (Pohl, 1989). *Objects* are class variables. Methods are referred to as *member functions*. In C++ 'everything is a data type'. Classes are simply new, user-defined data types with objects being variables of the specified type. In some ways this perspective on object-oriented programming parallels the Smalltalk philosophy of 'everything is an object'. Just as in Smalltalk everything – including basic elements such as integers – is an object, in C++ everything, from integers to complex ADTs, is a data type. This is a unique and very consistent perspective on OOPLs, and it is a natural and intuitive extension to the philosophy of the C language.

The class *and* struct *constructs*

In C++ a class is defined as follows:

```
class SomeClass: [private/public] ItsSuperClass (arguments)
{
public:      // public instance variables and member functions

protected:   // protected instance variables and member functions

private:     // this is the default, if no qualifier is specified
             // private instance variables and member functions
};
```

Let us analyze this further. The **class** keyword simply states that we are about to define a new ADT. The ':' indicates the immediate superclass. It can be omitted if **SomeClass** does not inherit from any other. The keywords in the square brackets immediately following the ':' indicate the degree of visibility of **ItsSuperClass** class members that **SomeClass** desires. We shall discuss this in greater detail later.

When a class is declared, the constructor of its superclass (all the way up the inheritance tree) is automatically invoked *before* its own constructor is invoked. For example, suppose we have the following class hierarchy (the notation **Class** ⇒ **SubClass** indicates that **SubClass** inherits from **Class**).

ClassA ⇒ ClassB ⇒ ClassC

Declaring a variable

ClassC instOfClassC;

will result in the constructors of **ClassA, ClassB** and **ClassC** being called in that sequence. The arguments in parentheses immediately following **ItsSuper-Class** are values with which the constructor of **ItsSuperClass** will be invoked when **SomeClass** is declared.

The data and member functions declared in the **private** section are visible *only* to the member functions of the class and **friend** functions (to be discussed), no other. It is not possible for the user to directly access them, nor are they visible to the derived subclasses. The **protected** members are visible *only to* member functions and derived subclasses. The **public** members are visible and accessible by everyone. In general, **private** members are for the hidden implementation details, while **public** members are for the interface to the users. Since **class** is an extension of **struct**, C++ maintains the same syntax for accessing class members. For example:

```
class Screen
{
public:
  Screen( );       // constructor
  ~Screen( );      // destructor
  void   Display (int xLoc, int yLoc);
  void   Undisplay( );
private:
  int    xCoord;
  int    yCoord;
};
main( )
{
  Screen screen;
  Screen *screenPtr = new Screen;
  screen.xCoord = 10;
  screenPtr → Display (5, 10);
  screen.Undisplay ( );
}
```

When defining a class, the same qualifiers can control the manner in which superclass members become visible within the class. If we have the following definition:

```
class ClassB : private ClassA { . . . };
class ClassB : ClassA         { . . . };
```

then **public** *and* **protected** members of **ClassA** become **private** in **ClassB**. This is the default, if no qualifier is specified. If however, we have the following:

```
class ClassB : public ClassA { . . . };
```

then:

1. **public** members of **ClassA** become **public** in **ClassB**; and
2. **protected** members of **ClassA** become **protected** in **ClassB**.

Recall that **private** members *cannot* be inherited in any case.

A **struct** is simply a special case of **class**. The difference is in the default visibility of the two. The variables in a **struct** are **public** by default and in a **class** they are **private** by default. A **struct** can have member functions, constructors and a destructor and can change their default visibility through the use of **private** or **protected**.

Multiple inheritance

This feature is available in C++ v2.0. The syntax is straightforward:

```
class ClassA :    [private/public] SuperClass1,
                  [private/public] SuperClass2,
                    . . .
{ . . . };
```

The parental relationship is described by a directed acyclic graph (Pohl, 1989). This graph is a structure with the classes being nodes and whose directed arcs point from base to derived class. It cannot be circular and so a class cannot inherit from itself.

Function overloading and **inline** *functions*

In a class, member functions with the same name can be overloaded to have different types and number of arguments as well as return types. This is used especially often with the constructor function. (There can however be *only one* destructor.) Constructors are used typically for storage allocation, object initialization and even conversion of one data type into another. Destructors are used for storage reclamation. If we define a new data type called **complex**, we would like to have the flexibility to

1. simply create a blank instance and set the values later, and/or
2. create an instance with specific values, and/or
3. convert another numeric type such as **int** or **float** to **complex**.

This can be achieved by overloading the constructor. For example:

```
class complex
{
public:
  complex( )                        // default constructor
  { real = 0.0; imag = 0.0; }

  complex(double r, double i = 0.0)   // use default argument
  { real = r; imag = i; }
private:
  double    real;
  double    imag;
};
```

In the definition of **complex** we find that we have included some function implementations within the class declaration. Such functions are automatically assumed to be **inline** and expanded by the compiler wherever they are accessed, thus eliminating function call overhead. The keyword **inline** is not required here.

Operator overloading

In C++ it is possible to overload standard operators such as '+' and '−'. This feature is useful in situations where operations between objects of a particular ADT has meaning. For example, in the case of the **complex** class defined above, it is logical and highly readable to have the following syntax:

```
complex    cplx1(1.0, 2.0);
complex    cplx2(5.0, 1.5);
complex    cplx3;
cplx3 = cplx1 + cplx2;
```

This is done using the **operator** keyword. For example

```
class complex
{
public:
  complex(double r, double i);
  . . .
  complex &    operator + (complex &);
  . . .
private:
  double       real;
  double       imag;
};
complex & complex::operator + (complex & cplx)
{
  real  += cplx.real;
  imag += cplx.imag;
  return *this;
}
```

The statement:

```
cplx3 = cplxl + cplx2;
```

will be translated to:

```
cplx3 = cplx1.operator + (cplx2);
```

Even though it appears that the '+' operator defined as above takes only one argument, it actually has two – the first is the implicit **this** pointer.

The ability to overload basic arithmetic operators can produce highly readable code; for example, redefining operators for matrices, vectors, and other data types. However, it can also be misused by applying it to unlikely situations. The general guideline is to apply it only to situations where the operator to be overloaded has an 'accepted' meaning with respect to the ADT in question.

Scope resolution operator

The scope resolution operator discussed earlier has greater significance when used with classes. It is used for accessing specific functions from a class, in the implementation of the member functions and for declaring **friend** functions. (Note that the scope resolution operator can be used with **struct** in the same manner as for **class**.) Following the convention of function prototype declaration, let us declare a class. The class definition is in the interface file (with a **.hpp** filetype) and the method implementations are in the implementation file (with a **.cpp** filetype).

```
filename: DialogWindow.hpp
  #include "Window.hpp"          // include superclass definition
  class DialogWindow : public Window
  {
  public:
    DialogWindow( );
    DialogWindow (int xLoc, int yLoc);
    ~DialogWindow( );
    void Display(int xLoc, int yLoc);
    void Undisplay( );
    void SetCallbacks( );
  // other members
  };
filename: DialogWindow.cpp
  #include "DialogWindow.hpp"    // include class definition
  DialogWindow : : DialogWindow( )
  {
    // implement constructor, create push buttons etc
  }
  DialogWindow : : DialogWindow(int xLoc, int yLoc) { . . . }

  DialogWindow : : ~DialogWindow( )
  {
    // reclaim storage allocated while this object was active
  }
  void DialogWindow : : Display(int xLoc, int yLoc) { . . . }
```

There is a major benefit to the developer as a result of this extension of the
scoping rules. While it is always better design to have small classes, with the
philosophy of 'one class – one function', it sometimes so happens that a class
becomes very big. This could be due to a variety of implementation and system-
dependent reasons that are beyond the control of the developer. For example,
when developing user interfaces, it may happen that maintaining and manipulating large amounts of context-sensitive information requires a class to have many
member functions. In such a situation, compiling the entire class at one time
could prove to be difficult due to limitations of the compiler. This could be
aggravated if the class requires a large number of #include statements. Using
the scope resolution operator, the class implementation can be split into smaller
files with references being resolved at link time. The following example should
help clarify this point.

filename: LargeClass.hpp

```
  #include    "SuperClass.hpp"
  class LargeClass : public SuperClass
  {
    // define instance variables . . .
    void        Method1 (. . .);
```

```
   . . .
   int          MethodN ( . . .);
   char         MethodNplus1 ( . . .);
   . . .
   double       MethodM ( . . .);
};
```

filename: LargeClassMethods1ToN.cpp

```
#include "LargeClass.hpp"
// include other class definitions that are required for the
// methods implemented here
void LargeClass : : Method1 ( . . .) { . . . }
. . .
int LargeClass : : MethodN ( . . .)   { . . . }
```

filename: LargeClassMethodsNToM.cpp

```
#include      "LargeClass.hpp"
// include other class definitions that are required for the
// methods implemented here
void LargeClass : : MethodNplus1 ( . . .) { . . . }
. . .
double LargeClass : : MethodM ( . . .)   { . . . }
```

The two **.cpp** files can be compiled individually and linked. This is far more efficient than having to compile the entire **LargeClass** at one shot. Even if the compiler were capable of handling large files, distributing the method implementations into more than one file encourages a degree of modularity that can enhance the development effort. If a change is made in **Method2** above, there is no need to recompile **MethodNplus1** through **MethodM**. For a large class, this offers the programmer the ability to make changes and quickly see their effect, instead of being slowed down by an unnecessarily large file. Of course, the desire to distribute the methods should be applied with caution.

The **this** *pointer*

The **this** keyword denotes an implicitly declared, unmodifiable, self-referential pointer. It can be thought of as a pointer to the object for which the function was invoked. It is a hidden argument to the function. The C++ compiler automatically declares a private instance variable:

CLASS_NAME * const this;

The **this** pointer can be compared to the variable '**self**' in Objective-C (discussed later).

friend *functions*

This is one of the more controversial features of C++. It is a means of breaking the cocoon surrounding the **private** variables of a class. A nonmember function that is granted access to the **private** section of a class is called a **friend** of the class (Stroustrup, 1987, p. 149). A function is made a **friend** of a class by a declaration in that class. (The class should *want* to have the nonmember function as a friend.) Examples:

```
class Screen
{
  friend void Graphic : : MoveGUIElement( );
  . . .
};
```

It is possible to make an entire class a **friend**:

```
class Window
{
  friend class GraphicManager;
  . . .
};
```

The **friend** function does not have a **this** pointer. The declaration can be placed in either the **private** or **public** sections of the class. While there is controversy regarding whether such a facility should be available at all, in commercial systems the need could well arise. A similarly controversial construct, the **goto** statement, has existed in all languages.

Good class design should reduce the need for this construct. However, situations where it could be useful are not hard to imagine. For example, suppose we have **ClassA** that is used by other classes but in **ClassB** it is used especially intensively. This could be for computation intensive tasks on data that is part of **ClassA**. One option is to make **public** those members of **ClassA** that are especially relevant to **ClassB**. But this would give access to all the other users of **ClassA**. In such a situation, good design would require that **ClassA** keep its **private** data and declare **ClassB** (or relevant methods from **ClassB**) to be **friends**. One piece of advice is to use it sparingly.

virtual *functions*

Virtual functions help in implementing (a partial degree of) dynamic binding in C++. When a virtual function is accessed, the function selected depends on the class of the object being accessed, not on the pointer type. It is dependant on the fact that pointers to derived classes are compatible with pointers to parent classes (Wiener and Pinson, 1988, p146). This means that a pointer to a parent class can point to either a parent class object or at instances of classes derived from it.

Suppose we had a function with the same name, arguments and return type in both the parent and derived classes (an overloaded function). We would like the function from the appropriate class to be invoked based on the object obtained at run time. This requires that the overloaded function be declared as **virtual** in the parent class only.

As an example, suppose we had a class hierarchy arranged as follows:

BaseClass ⇒ SubClass ⇒ SubSubClass

and each implemented a method called **dumpInternals( )**. Based on some error condition, we would like the object being pointed at to 'dump its internals' on the screen. But since we do not know what conditions will exist at the time of the error, we would like to bypass the static typing of the pointer. This can be achieved in the following manner.

```
class BaseClass
{
public:
  int       BaseAction( );
  virtual void dumpInternals( );
  . . .
};
class SubClass : public BaseClass
{
public:
  int SubAction( );
  virtual void dumpInternals( );
  . . .
};
class SubSubClass : public SubClass
{
public:
  int       SubSubAction( );
  void      dumpInternals( );
  . . .
};
void errorHandler (BaseClass & offendingObject)
{
  offendingObject.dumpInternals ( );
}
main ( )
{
  int              error;
  BaseClass        base;
  SubClass         sub;
  SubSubClass  subsub;
  error = base.BaseAction ( );
  if (error) errorHandler (base);

  error = sub.SubAction( );
  if (error) errorHandler (sub);

  error = subsub.SubSubAction ( );
  if (error) errorHandler (subsub);
}
```

Another example application is through the use of vectors:

```
BaseClass   vector[10];
```

Each element in **vector** could be an object of type **BaseClass**, **SubClass** or **SubSubClass**. In order make each one dump its internals:

```
for (int i = 0; i < 10; ++ i)
    vector[i].dumpInternals( );
```

For developers familiar with dynamic binding in other languages such as Smalltalk or Objective-C, this is a laborious and contrived way of achieving it It is however, consistent with the original intention of making C++ an efficient, strongly typed language – with the provision for dynamic type checking being available only in specific cases where it is needed. C++ is designed to be statically typed and to provide stronger type checking than C.

Smalltalk and Objective-C developers will notice some limitations. The dynamic type checking is allowed only along a single branch of the inheritance tree. This is very different from Objective-C where dynamic binding is provided across the entire inheritance hierarchy. Another limitation is that the dynamic binding is available only from the class on down the branch. An example will help clarify this point. Suppose we changed the example above as follows:

```
class SubSubClass : public SubClass
{
public:
  int SubSubAction( );
  virtual void dumpInternals( );
  . . .
};
class SubSubSubClass : public SubSubClass
{
public:
  int SubSubAction( );
  void dumpInternals( );
  . . .
};
void errorHandler (SubSubClass & offendingObject)
{
  offendingObject.dumpInternals( );
}
```

In this situation, a call to **errorHandler( )** with an object of type **BaseClass** or **SubClass** will result in an error. This can be understood by considering the fact that a pointer to a derived class *cannot* point to a parent class object. In other words, a derived class pointer is compatible with its parent class but not the other way around.

7.5.3 Summary

This is an overview of some of the features available in C++ for object-oriented programming. It is not a complete list (interested readers are further referred to Ellis and Stroustrup (1990) for up-to-date discussions of other features). It is important to note that C++ is still evolving. New features are being introduced and until an ANSI standard is established, it is hard to predict when the evolution will stop. Many of the extensions under consideration are exotic and designed to handle very special cases. For example, it may soon be possible for a class to derive from its parent in **protected** mode (in addition to **private** and **public**).

Programmers will find the syntax of C++ much harder to read in comparison to Objective-C or Smalltalk. In addition, even though the fundamental concepts underlying C++ are intuitive (such as the extensions to **struct** and the uniform view of data types), the innumerable 'micro' enhancements and special cases have turned it into a language that is (especially for beginners) complex and hard to understand. However, if software managers ensure that the use of the 'micro' features is minimized, they will find that C++ offers most of the benefits of an OOPL combined with the speed and efficiency of C.

7.6 Objective-C

Objective-C is provided by Stepstone Corporation and is the development environment on the NeXT computer. It is the more conservative of the two (i.e. Objective-C and C++) in terms of the extensions to the C language. It is based on the ANSI version of C and recognizes all standard C syntax.

Objective-C was developed by C programmers who were looking at the advantages of OOP for conventional projects (Tello, 1989). It is strongly influenced by Smalltalk and is really only a set of Smalltalk-like features provided on top of the base C language. However, not every Smalltalk features is present in Objective-C. The terminology is quite similar and Smalltalk users will find it quite natural to migrate to Objective-C. It will be useful to briefly review some basic Objective-C terminology; see also Meyer (1988) and Adiga and Gadre (1990).

An '*object*' is a bundle of data and procedures that act on that data. The data are stored in '*instance variables*' that are private to the object and accessible only by it. The procedures are called '*methods*' and are accessible by any other object. A '*class*' is a static concept, a piece of program text (Meyer, 1988, p. 94). and is a means of defining an object in terms of its abstract type. Objects are '*instances*' of their respective class.

7.6.1 Keywords

The syntax of Objective-C is simple, transparent and straightforward. Objective-C reserves for itself a set of six keywords, each of which is preceded by the symbol '@'. These keywords are:

1. @interface
2. @implementation
3. @end
4. @public
5. @selector
6. @defs

In addition, Objective-C makes the following additions to the C language:

1. one new type (**id**);
2. one new expression (for messages: [**receiverObject message**]);
3. one new constant type (the selector constant).

It is important to note that these are not extensions to the C definition itself but are similar to preprocessor directives, which are translated into C. This is because Objective-C adds object-oriented capabilities to C and does not attempt to re-define it.

Objective-C comes standard with a mature class library called ICpak 101 (Stepstone, 1989). As a result, the initial hurdle of developing one's own fully debugged set of foundation classes is eliminated. While this library is inspired by Smalltalk classes, it is not as extensive as the set of libraries provided as part of that environment. However, the ICpak 101 classes are extremely well thought-out and designed. It is possible for the user to come up to speed quite quickly in developing new applications.

7.6.2. Messages and selectors

Messaging in Objective-C is done through a single, highly optimized messager. The use of a single messager allows highly informative stack backtraces to be presented to the developer in the case of an error.

However, the programmer also has the option of overriding it by statically binding messages to objects. There is a significant benefit to the Objective-C approach. The *user* of a class has the option of deciding which objects and methods to bind statically. It is *not pre-decided by the developer*. This is significant because the developer of a software-IC cannot foresee all possible uses of his/her class. If the developer makes decisions for the user then it limits the user's options. With 'consumer-side' decisions, all the options are with the user, without in any way weakening the developer's choices.

According to Cox (1986), messages are a modularity mechanism. Messaging creates the encapsulation of data and procedures that is called an object. A message expression behaves similarly to an ordinary function call with the extra feature of a selection mechanism that results in dynamic binding. Functions and messages differ in other respects as well.

Functions could have zero or more arguments while messages have at least one – an argument that identifies the object that is to receive the message. In Objective-C this implicit argument is identified by the reserved keyword 'self' while in C++ it is the 'this' pointer. (Both are further discussed in other sections of this chapter.) In object-oriented programming the piece of code that is executed in response to a message is dependent on the object receiving the message. The part of a message that tells the receiver of the message which piece of function code to execute is called the *message selector* (Cox, 1986). According to Cox, the only significant difference between conventional programming and object-oriented programming is the selection mechanism. It allows the responsibility of identifying the exact block of code to be executed to be moved from the consumer to the supplier of the code.

In Objective-C all objects translate selectors to implementations via a single function called the *messager* and all message expressions compile into calls on it. This function is called _msg ( ). The arguments to this function are the id of the receiver and the message selector. The mapping between the selector and a pointer to the appropriate function is maintained in '*dispatch tables*'. Each object contributes a dispatch table that the messager accesses.

Objective-C provides the '@selector (. . .)' construct which can convert arbitrary C strings to the appropriate selector code. This allows 'computed messaging' where even the message to be sent is not known until runtime.

7.6.3 Classes, methods and instance variables

Class definitions are made in two parts, the *interface* and the *implementation*. It is recommended that the interface be placed in a file of type '.h' while the implementation is in a file of type '.m'. Both have a filename which is the same as the class name. The interface file (file-type '.h') contains the class definition and the method prototypes. The implementation file (filetype '.m') includes the interface file and contains the method implementations.

There is a distinction between the types of methods in a class – '*factory methods*' and '*instance methods.*' Factory methods are accessible only *at the time of* object instantiation and are identified in the source code by the symbol '+'. Instance variables are accessible only *after* instantiation and are identified by the symbol '-'.

The following is an example:

```
filename: MyClass.h
  #import "SuperClass.h"

  @interface MyClass : SuperClass
  {
          char instanceName [32];
          // other instance variables
  }
  // Factory Methods
  + create;
```

```
...
// Instance Methods
- (char *) instanceName;
- instanceName: (char *) aName;
...
@end
```

filename: MyClass.m

```
#import "MyClass.h"
@implementation MyClass
+ create
{
        self = [self new];
        strcpy (instanceName, " ");
        // initialize other instance variables...
        return self;
}
- (char *) instanceName { return instanceName; }
- instanceName: (char *) aName
{
        strcpy (instanceName, aName);
        return self;
}
...
@end
```

The above example introduced Objective-C concepts and syntax that will be explained in the following paragraphs. The variable types that are supported include all the basic C types and **id** – a new data type defined in Objective-C to refer to variables that point to dynamically allocated and typed objects. The default return type from methods is **id**.

The keyword **@interface** tells the compiler that the class interface follows. The name of the class is followed by the super class name. **@implementation** signifies the start of the implementation section. The keyword **@end** signals the end of the appropriate section.

Methods and Messages

The message syntax is as follows:

```
[receiverObject doSomethingWith:parameter];
```

A message expression consists of a pair of square brackets [. . .] enclosing a receiver, a selector and optional arguments (Stepstone, 1989). The receiver is any valid Objective-C expression that evaluates to a valid object **id** at runtime. Hence it is possible to nest messages as follows:

```
[objectA doSomethingWith:[objectB returnSomeValue]];
```

A selector represents the method to be executed by the receiver object and is created where the appropriate method is defined. It is referenced by its name, return types and argument types.

Unless specified, the default return type from a method and the type of any parameters passed into a method is **id**. Using the ':' symbol, it is possible to pass more than one parameter in a message and make the messages more readable:

[object task : "display" xCoord : 100 yCoord : 50];

There are three types of messages:

1. unary messages (such as '**create**' above);
2. keyword messages (such as '**task : xCoord : yCoord :**' above);
3. variable argument messages (for example [**receiver errmsg:** "error type = %d, message = %s", type, msg];).

Methods are accessed by messaging, not by name as for C functions (Stepstone, 1989). The methods are assigned generated names (by the Objective-C compiler) and declared **static**, so they have only file scope. Instance variable names are in scope only within method implementations.

Implicit method arguments

Every method automatically receives the formal parameters **self** and _cmd. **self** is of type **id** and identifies the receiver of the message that activated this method. **self** is a pointer to the private part of the object. (Readers may compare **self** to the **this** pointer in C++.) It is a standard C function argument and so can be reassigned. _cmd is of type (**char ***) and is the selector for this method.

7.6.4 Dynamic binding

Objective-C supports true, run-time, dynamic binding for variables of type **id**. This means that the class of the object to which the message is being sent is resolved only at run time. Objective-C provides this powerful feature in an ANSI C environment, which is statically typed. This is the default in Objective-C. This feature makes it very easy to quickly prototype system behavior. Dynamic binding is made possible through the use of the new type **id** and the selector mechanism.

An object consists of two parts: a 'private' section that contains only the data that is private to the object and a 'shared' section that is shared by all instances of that class (such as methods). These sections are managed through the definition of two **struct**'s – _PRIVATE and _SHARED. The definition of _PRIVATE depends on the object class, but the very first field is always a pointer to its shared part. This variable is always called '**isa**'. It will be useful to look at the definition of this **struct**.

```
struct _SHARED
{
  struct _SHARED          *isa;
  struct _SHARED          *clsSuper;
  char                    *clsName;
  char                    *clsTypes;
  short                   clsSizInstance;
  short                   clsSizDict;
  struct _SLT             *clsDispTable;
  long                    clsStatus;
  struct modDescriptor    *clsMod;
};
```

Let us look at some of these variables. For a complete explanation the reader is referred to (Cox, 1986) and (Stepstone, 1989). Every object holds a dispatch table of methods (**clsDispTable**) and a pointer to the shared part of its super class (**clsSuper**) from which it inherits additional methods. The dispatch tables are not declared within the shared structures but are referenced through a pointer. The **clsTypes** string holds the types of each of the instance variables using a code for each type. For a description of the codes used, the reader is referred to the Objective-C Compiler Reference Manual (Stepstone, 1989). This information is particularly useful for dumping objects to data files and recreating them at a later point in time – i.e. passivation/activation (Cox, 1986).

The variable type **id** is defined in a **typedef** statement that is generated in each class file. The specific statement is

```
typedef struct _PRIVATE *id;
```

The definition of **_PRIVATE** will vary for each class. It contains the values of the private section of an object, namely the instance variables. For example, in the following situation:

```
@interface Class : Object
{
    char name[32];
}
```

the definition of **_PRIVATE** will be as follows:

```
struct _PRIVATE
{
  struct _SHARED    *isa;
  unsigned short    attr;
  unsigned short    objID;
  char              name[32];
};
```

The first three variables, **isa**, **attr** and **objID**, are inherited from **Object** (Stepstone, 1989). The **Object** class is defined in Objective-C and is part of the

Runtime Library. **Object** is at the root of the inheritance tree. The name variable is defined in **Class**.

Computed messaging

In normal messaging the class of the receiver is unknown at compile time but the message is. It is possible to delay binding even further, by using *computed* messaging where even the message to be sent is determined only at runtime. This is achieved through the use of the '**perform:**' message (defined in the **Object** class) and the **@selector** construct. For example, if we have:

```
@interface Class : Object
{
    . . .
    + task;          // factory method
}
```

then we could do either of the following:

```
[Class task];
[Class perform : "task"];
[Class perform : @selector (task)];
```

It is fastest from an execution speed point of view to use the first technique and slowest using the second. However, the second technique allows the programmer to send messages based on C strings that may be read in from a file or even typed in by a user at run time. In addition, the C function '**_cvtToSel( )**' translates strings to selector codes. For example:

```
char *msgName = (*_cvtToSel) ("anotherTask : withParam : ");
```

returns the selector code for the message anotherTask:withParam: or NULL if it does not exist in this application.

There is a risk in this. There must be a way to check if the computed message being sent is valid for this object or not. Objective-C provides such a facility through the message '**respondsTo:**'. For example:

```
if ([Class respondsTo : @selector (task)])
    [Class task];
```

These are extremely powerful features that allow the development of dynamic, efficient and robust applications.

7.6.5 Static binding

Objective-C allows the developer the option of statically typing objects for efficiency purposes. The constructs that support this facility are as follows:

1. **@defs**
2. **@public**

3. direct method handles (using 'methodFor : ' and 'instanceMethodFor : ')
4. compiler option '-sBind'

For the sake of explanation, suppose we had the following hierarchy of classes (GUI, graphical user interface):

```
@interface GUIElement : Object        // GUIElement Class
{
  int      xLoc;
  int      yLoc;
}
@end

@interface Window : GUIElement         // Window Class
{
  int      height;
  int      width;
@public
  BOOL isMapped;
}
+ new;
– moveToX : (int)x Y : (int)y;
– (int) width;
@end
```

The @defs *construct*

The **@defs** construct generates the internal data structure of its argument. So the statement:

```
struct _guiElement {@defs (GUIElement)} *guiElement;
```

generates the code:

```
struct _guiElement
{
  struct _SHARED    *isa;
  unsigned short    attr;
  unsigned short    objID;
  int               xLoc;
  int               yLoc;
} *guiElement;
```

This is the same definition which methods in the GUIElement class will see. The programmer can now directly access the instance variables of an instance of GUIElement. For example:

```
guiElement = aGUIElement;
x = guiElement → xLoc;
y = guiElement → yLoc;
```

The @public *construct*

For statically typed objects, the variables in the @public section of the class definition can be directly accessed using the **struct** notation. For example:

```
Window aWindow;
if (aWindow → isMapped) { . . . }
```

Any similar attempt to access private instance variables is not permitted.

Direct method handles

Direct method handles permit the programmer to obtain a pointer to the function which implements a message for a given instance. For example:

```
id (*funcPtr) ( );
funcPtr = [Window methodFor : @selector(new)];
```

will set **funcPtr** to point to the function that will be activated if **Window** were sent the message new. However, the user now has to supply the arguments to this function when it is called:

```
(*funcPtr) (Window, @selector (new));
```

If we are dealing with objects and instance methods:

```
funcPtr = [aWindow instanceMethodFor : @selector (moveToX : Y : )];
(*funcPtr) (aWindow, @selector (moveToX : Y : ), 10, 20);
```

If we are using this facility within a method, substitute 'self' for 'aWindow'. It must be remembered that the programmer be careful in the use of this facility. The Objective-C Compiler Reference (Stepstone, 1989) has further details on avoiding errors.

In the above example it was assumed that the function returned type id. However if it returned some other type, then typecasting will be necessary. For example:

```
int    windowWidth;
int (*funcPtr) ( ) = (int (*) ( )) [self
                        instanceMethodFor : @selector (width)];
windoWidth = (*funcPtr) (self, @selector (width));
```

This facility defeats the purpose of dynamic binding and is prone to error. It must be used only for repeated and compute intensive operations on instances of the same class.

The -sBind *compiler option*

The option -sBind informs the Objective-C compiler to statically bind all messages to statically typed objects.

It must be noted that with these features programmers have the ability to decide which methods are statically bound *when they use* the classes and methods. It is not decided by the supplier of the code. As a result, *users* of the software-ICs have the option of deciding whether to optimize their code or not. *The decision is not made for them in advance.*

7.6.6 Class posing

This is another powerful feature of Objective-C which allows classes to *pose as* their immediate superclass at runtime, in effect becoming that class for all practical purposes. This is done to correct defects in the superclass or to augment its functionality without tinkering with its source code. It also allows the incremental compilation of very large classes.

If the implementation of a very large class (called **LargeClass**) was distributed between **LargeClass** and its subclasses **SubClassA** through **SubClassN**, then the effect of incremental compilation could be achieved by:

```
@interface SubClassI : LargeClass
{
  // NO NEW INSTANCE VARIABLES ALLOWED
}
+ initialize
{
  [self poseAs : LargeClass];
  return self;
}
```

This will result in all messages being sent to **LargeClass**. The limitations to this are that the classes may only pose as their immediate superclass and posing classes may only define or redefine methods; they may not add new instance variables.

7.6.7 Multiple inheritance

Objective-C *does not support* multiple inheritance. The reasons for this are laid out in detail by Cox (1986). In Objective-C, superclass methods are not copied into the subclasses but are accessed via the messager. With single inheritance, the instance variables are laid out sequentially in memory, so methods using them can be bound to static offsets that are known at compile time. If, however, we have multiple inheritance, these offsets will no longer be valid in a class that inherits from many different classes.

This means that in order for the dynamic lookup mechanism in the messager to work correctly, the methods from each superclass will have to be copied into the subclass and recompiled with the (now changed) offsets. As a result, the compiler will have to know about both the interface *and* the implementation of the associated superclasses, when it attempts to compile a class.

While this is quite feasible from a compilation point of view, it creates problems that go against some of the underlying philosophies of Objective-C. Firstly, it creates an unnecessary file management problem. Secondly, it is now necessary to know the internals of a software-IC, since it is impossible to compile its subclasses by knowing only the interface (or 'specs'). Thirdly, it is impossible to distribute code in binary form. One has to supply the source code as well. The combination of these factors made Cox decide against providing multiple inheritance.

7.6.8 Passivation/activation

For a given process, its address space determines its range of identifier uniqueness. While a user may have more than one process running, each is isolated from the other except for information passed via pipes and filters. Objective-C provides an automatic passivation/activation mechanism whereby information can be passed between processes, even those running on different computer systems. This mechanism converts objects into a symbolic representation in a permanent file that can be transferred between processes.

'*Passivation*' builds a symbolic representation of any object and '*activation*' reconstructs the object from this representation (Cox, 1986). The methods that support this powerful feature are '**storeOn: (char *) fileName**' and '**readFrom: (char *) fileName**'. Any object can be sent the **storeOn:** message while sending any factory object the **readFrom:** message that will restore it. When a saved object is restored it may have to do some of its own reinitialization. This is implemented in a method called '**awake**' and every object is automatically sent the **awake** message once it is reconstructed. The user may implement the **awake** method in any application-specific manner.

When an object is passivated, its instance variable types are stored in a coded form. This code is the same as the one used for the **clsTypes** variable (see above). While this is quite extensive, there are limitations which the user must be aware of. For a discussion of this, the reader is referred to Cox (1986, p.128).

7.6.9 The Objective-C Runtime Library

The Runtime Library is a set of C functions, include files and classes. For a complete description of these elements the reader is referred to Stepstone (1989). The Runtime Library provides foundation support for the Objective-C environment. It contains definitions of the functions and constructs we have discussed above, such as **cvtToSel ( )** and so on. The class hierarchy is as follows:

```
Nil        Object
           AsciiFiler
           ObjGraph
           Unknown
```

7.6.10 The ICpak 101 library

Objective-C comes standard with a mature library of classes which provide the user with many powerful features. Using these classes enables the user to come up to speed very quickly with the language and start developing non-trivial applications.

The library classes perform many basic functions such as dynamically creating new objects, creating arrays of integers and objects and maintaining linked lists. This encourages the user to concentrate on applying object-oriented design principles and follow the paradigm of 'send messages to objects' rather than the traditional C philosophy of manually navigating pointers. In the author's experience, it is possible to write highly readable, non-trivial applications in Objective-C without requiring the explicit manipulation of object pointers.

The ICpak 101 classes are arranged in the following hierarchy (from the Objective-C ICpack 101 v4.0 User Reference Manual, Stepstone, 1989). Recall that the **Object** class is part of the Runtime Library and not ICpak 101.

```
Object
  Array
    IdArray
    IntArray
  Assoc
  BalNode
    SortCltn
  ClassTester
  Cltn (Collection)
    OrdCltn
    Stack
    Set
        Dictionary
  IpSequence
    Sequence
  Point
  Rectangle
  String
```

7.6.11 Summary

The above discussions demonstrate that Objective-C is a powerful object-oriented programming language. Large and efficient object-oriented applications can be developed using it. In addition, the combination of the following factors:

1. simplicity and readability of syntax and extensions;
2. true dynamic (as well as static) binding;
3. consumer side options;
4. mature foundation classes; and
5. facilities such as passivation and activation,

make Objective-C the ideal choice for development environments and for those environments where there is a rapid turnover of developers (such as in a research group at a university; see, for example, Glassey and Adiga, (1990). Objective-C is also preferrable in situations where a transition to object-oriented programming is being made and the developers are not familiar with OOP.

7.7 Comparing Objective-C and C++

Having discussed the specific features of these two major object-oriented dialects of C, it will be useful to compare and contrast them.

7.7.1 Criteria

Objective-C and C++ will be compared in the context of certain important considerations such as information hiding and dynamic binding. (See also Stroustrup (1990) for additional viewpoints.)

Scoping

In C++ there are three kinds of scope: local, file and class. Local scope applies to variables that are declared within a block of code. File scope applies to variables defined at the top of a source file, outside of any block or class. They are available to all implementations defined after their declaration in the same file. Class scope refers to variables defined as instance variables of a class. In Objective-C however, we have only local and file scope, which have been available in C.

The scope resolution operator (': :') in C++ allows the implementations of member functions to be distributed across source files with the operator being used to resolve references. As explained earlier, this also allows the breakup of large classes into smaller files which can be individually compiled and linked.

Class posing

Objective-C provides the facility for a class to *pose as* its immediate superclass. This allows the user to augment the superclass or correct its deficiencies without tinkering with its internals. This is another result of the Objective-C philosophy of providing as many '*consumer-side*' options as possible. C++ does not provide this or any other similar facility. Class posing allows the piecewise compilation of large classes.

Information hiding

In Objective-C the options available to the programmer for information hiding are simple and restricted. All instance variables are inaccessible to the outside world. All methods are fully accessible. While this is easy to comprehend and ensures that a standard of some sort will exist across all classes, it fails to anticipate the different requirements that may surface in a commercial development environment. In particular, the full accessibility of *all* instance methods (factory methods if we are considering initialization) to the outside world can potentially undo any benefits of data hiding. Consider, for example, methods that are implemented solely for the sake of manipulating internal data. This scenario is very plausible and could happen, for instance, with a block of code that is used in different methods. This method is designed to be used solely by other methods within the class. However, there is *absolutely no way* of preventing another user from carelessly using it, thus directly interfering with the object's internal data and destroying its integrity.

In C++ however, there are different levels of opacity **private, protected** and **public**. As a result, all internal data and housekeeping methods can be **private**, thus sealing them off from the rest of the world. Only those methods that are required by the users can be made **public**. Even derived classes cannot access **private** data or methods, thus eliminating the possibility of derived classes causing errors through the careless use of superclass methods and data.

This facility is particularly useful in development environments with large numbers of programmers. By allowing access only to those elements that have mutually been accepted as the class interface, the possibility of misuse is reduced.

Interface to subclasses

In Objective-C the interface of a class to its subclasses is again quite simple. All instance variables of a class are inherited by its subclasses and are invisible to the outside. All factory methods become factory methods in the subclass and the instance methods get inherited as well. All methods are fully visible to the outside (depending on whether it is at the time of initialization or not). While this is consistent and easy to comprehend, it opens up the possibility of a subclass destroying the integrity of the class simply by inheriting from it and manipulating its internals.

In C++ a subclass can inherit in **private** or **public** mode. As a result a developer of a class (say **ClassA**) can block all access to its superclass by inheriting from it in **private** mode. In this way, the **protected** and **public** members of the superclass become **private** in **ClassA** and are hidden from the subclasses of **ClassA** as well as the outside. (Of course, if **ClassB** inherited from the superclass of **ClassA** in **public** mode, the subclasses of **ClassB** will have full access to the visible members of the superclass.) On the other hand, by inheriting in **public** mode, the visible members of the superclass remain so in its subclass.

While this may be more confusing than the straightforward inheritance mechanism of Objective-C, it does offer developers a wider choice in interfacing to subclasses. In large systems, exceptions always occur and greater flexibility is preferable.

Multiple inheritance is not available in Objective-C. It is available in the AT&T C++ v2.0 compiler (Ellis and Stroustrup, 1990).

Dynamic binding and polymorphism

As we have already seen, Objective-C offers true dynamic binding across the entire inheritance tree. This feature is the default and requires no extra work by the programmer. It is simple and straightforward to both understand and use. And it allows and encourages the rapid development of code.

Objective-C also supports static binding. Constructs such as **@public** and **@defs** allow the programmer to access instance variables directly if need be. Direct method handles provide access to the function referenced by the selector. These help eliminate the overhead associated with dynamic binding. The crucial benefit is that the option of using static binding is in the hands of the user. It is not decided in advance by the developer. This 'consumer-side' decision-making frees the supplier from having to change the source code to support extra requirements by the user.

C++, on the other hand, offers only a limited version of dynamic binding. As explained earlier, it is allowed only along a specific branch of the inheritance tree. It also requires a significant amount of extra effort. In particular, it forces the developer to know and decide in advance, what methods should or should not be '**virtual**'. This 'supply-side' decision-making inhibits flexibility. In addition, while static typing might lead to more efficient code, it takes more time and effort to develop it.

Apart from the more pleasant syntax and ease of use, these major features could well tilt the balance in favor of Objective-C, depending on the applications and environment it is to be used for. For artificial intelligence applications dynamic binding is a critical requirement (Tello, 1989). In addition, the availability of true dynamic binding in a C compiler environment means that the user has the option of statically typing and using low-level C constructs for enhanced speed and efficiency *wherever required*, without giving up any benefits of dynamic binding. In C++, using dynamic binding is far more unnatural and involved.

Design and prototyping

For most commercial software systems it is impossible to completely and accurately specify system details and behavior from the very beginning. Almost always, one learns and improves through the implementation of prototypes. Then, in an incremental manner, the prototype is extended to provide the full functionality that is required. This approach is formalized by Boehm (1988). It is also essential that designs be converted into implementations with a minimum of fuss. As a result, the implementation language of choice should provide the developer with facilities for rapid prototyping as well as ease of migration from design to implementation. Dynamic binding, polymorphism, reusability and modularity as well as consumer-side options facilitate both these requirements.

It is important to understand the usefulness of consumer-side options. In a development group where the high-level system requirements are understood but the mid-to low-level functionality is still to be fully crystallized, it is very limiting to depend on the *supplier* of a software-IC to provide features such as dynamic type checking. It is far more effective for the *user* to decide how and when to use such features. The same applies to migrating from design to implementation. When different developers are working on different aspects of a design, it is frustrating to find that in order to satisfy one's own requirements of dynamic/static typing, one has to go to the supplier of a software-IC to reopen and redefine it. It is far more efficient to make such decisions *at the time of use*.

From these considerations it is obvious that Objective-C is the more powerful language in terms of migrating from design to implementation as well as proto-typing. The ability to decide at the time of use whether one wants to dynamically or statically type variables and bind methods is available in Objective-C. In C++ such decisions have to be made by the supplier, and that too in an unnatural manner. In addition, the ability to pose as a superclass in Objective-C allows a consumer to augment a class without having to open it up.

Other Issues

C++ is the more efficient language of the two. It retains the flavor of C and does not attempt in any way to reduce the manual navigation and manipulation of pointers that is associated with it. This may not be a major problem for pro-grammers already proficient in C. But for anyone wanting to migrate to a C based OOPL, C++ is complex, confusing and a little too much to swallow. Objective-C on the other hand is far simpler to comprehend. The syntax is much easier on the eye – a view supported by other reviews of the language; for example, Tello (1989). And with the foundation library, it is quite possible to eliminate the nitty-gritties of C.

In addition, Objective-C supports true dynamic binding. This makes it a far superior tool for rapid development. In many situations, development time is far more critical than the low level optimization of every data structure or function. This is becoming more so with the advent of high performance computing equip-ment. It is more important to improve software productivity than to exclusively focus on hardware performance. According to Shemer (1987), the hardware cost of systems has *dropped* from 90% in the 1950s to 10% in the 1990s, with the relative cost of software systems *increasing* at a similar rate.

The definition of Objective-C is proprietary to Stepstone and is not in the public domain. This is one of the most serious liabilities of the language and one that has contributed more than anything else to its failure to penetrate commercial applications with the success of C++. Currently there are many software vendors supplying C++ programming environments. These include source level compilers and debuggers as well as CASE tools and class browsers. Support of this kind is not available for Objective-C. In particular, the absence of native compilers and debuggers is a serious liability.

7.7.2 Summary

The superior OOP capabilities of Objective-C such as dynamic binding, and a variety of consumer-side options makes it an ideal choice for prototyping. Applications such as those related to artificial intelligence may decide that these are crucial enough to make Objective-C the production language as well. However, for programming large scale systems, the support that is available for C++ (in the form of compilers, debuggers and browsers) as well as its features (such as the scoping operator and default function arguments) make it the preferred language. In addition, since the language definition is not proprietary to any one organization, there are many different vendors and products to choose from.

Harmony is usually found in the middle of two extremes. The ideal progression would be to use Objective-C to implement the design and prototype the system, since it is ideally suited for that. Once the developers have had a chance to work out the nuances of the system and establish the inheritance hierarchy and message protocols, it will be obvious where and how static and dynamic binding is to be used. This is an ideal time to move to C++. Since the skeletal structure of the system has been established, having only supplier-side options will not be too much of a liability. With the support of debugging tools, C++ can be used to fine tune the final system.

7.8 Other object-oriented programming languages

We now briefly discuss some of the other object-oriented programming languages that are available. Interested readers are also referred to Rumbaugh *et al.* (1991).

7.8.1 MPW PASCAL

MPW PASCAL is a hybrid object-oriented programming language. It is built on top of PASCAL in a manner similar to C++ and Objective-C. MPW PASCAL is a statically typed OOPL and is similar in some ways to C++. For example:

```
GUIElement = OBJECT
  name : STRING(16);
  xLoc : INTEGER;
  yLoc : INTEGER;
  PROCEDURE moveToLoc (x: INTEGER, y: INTEGER) : INTEGER;
END;

Window = OBJECT (GUIElement)
  ptrLocX : INTEGER;
```

```
    ptrLocY : INTEGER;
    PROCEDURE moveToLoc (x: INTEGER, y: INTEGER) : INTEGER;
    OVERRIDE;
END;
```

The new procedure is used to initialize instances. Variables of these types are declared in the same manner as variables of standard PASCAL types. For example:

```
VAR
    aGUIElement : GUIElement;
    aWindow : Window;
```

Messages are sent in the following manner:

```
    aWindow.moveToLoc (10, 20);
```

Recall that in C++, pointers of a class are compatible with those of its super-class (but not the other way around), with this fact being used to implement dynamic binding. A similar concept applies to MPW PASCAL. For example, **aGUIElement** could hold either **GUIElement** or **Window**. Each would behave appropriately on being sent the **moveToLoc** message.

The advantage of this language is that it extends PASCAL to provide object-oriented capabilities. Its benefits to existing PASCAL-based systems can be significant.

7.8.2 ACTOR

Actor languages could be defined as concurrent 'computational agents' which react in a declarative or procedural manner to incoming messages (Saunders, 1989). The commercial product ACTOR, from the Whitewater Group, operates on the IBM PC and is integrated with the Microsoft Windows environment. Despite its name, ACTOR resembles Smalltalk more closely than the earlier MIT group of actor languages.

ACTOR was introduced in early 1987 as an interactive, object-oriented pro-gramming language (Guttman, 1988) designed to enhance the capabilities as well as the usability of the Microsoft Windows Software Development Kit (SDK). Among the many ACTOR features that are not available in the SDK is an object library (see also Tello (1989) and Crabb (1990)).

ACTOR is an interpreted language making it ideal for prototyping and testing concepts 'on the fly'. It is economical in its use of machine resources. It is fully object-oriented in the same sense as Smalltalk; even fundamental data types are implemented as classes. It provides only single inheritance. ACTOR is recommended as an introductory and/or prototyping language for Windows-based applications as well as a general introduction to object-oriented concepts and programming.

However, encapsulation in ACTOR is compromised in the sense that all in-stance variables of an object are freely modifiable by any other object. The user loses one of the major benefits of OOP, namely data hiding. It is difficult to understand how this can be a benefit. On the one hand it forces everything to be an object and on the other allows unrestricted access to all data members. Developers seeking the full range of benefits of using an OOPL need to be aware of this aspect of this language.

7.8.3 Eiffel

This language was released in 1987 by Bertrand Meyer of Interactive Software Engineering, Inc. While it is implemented as a C code generator, the influence of C is not noticeable. It provides more compile-time type checking than other OOPLs. Type checking follows inheritance rules (Rettig *et al.*, 1989). It also provides continuous run time garbage collection – another plus for a compiled OOPL. It runs on UNIX systems and provides many run-time debugging tools. The graphical user interface is based on X-Windows which is another plus since it enhances portability. Like the two languages above, a drawback of Eiffel is that the language definition is proprietary. This restricts the growth of the language as well as the options available to the developer.

7.8.4 Trellis

The Trellis system has been developed by the Digital Equipment Corporation (DEC). It is an integrated programming language and environment that provides tools and mechanisms necessary for developing object-oriented programs (Kilian, 1990). The browsing tools enable the programmer to browse Trellis types and the operations they contain. The incremental compiler is capable of compiling even a single operation. It compiles Trellis into machine code.

Trellis provides strong type checking and multiple inheritance. Methods are called '*operations*' and '*class*' and '*type*' are used interchangeably. An operation can be '*private*' or '*public*'. Those operations that are visible only to subclasses are '*sub-type visible*' (Gallagher, 1991). This is quite similar to the **private**, **protected** and **public** visibility in C++. Trellis also provides the programmer with exception handling capabilities.

In order to use libraries written in languages other that Trellis, '*call in*' and '*call-out*' libraries are provided. These facilitate (for example) the conversion of Trellis objects into representations understandable by other languages. In general, Trellis is a powerful, tightly integrated language and environment that supports extension and customization.

7.9 Summary and conclusion

In this chapter we have summarized the important features and characteristics of different object-oriented programming languages. We find that the majority of the object-oriented programming languages available today are extensions to either LISP or C. The reader will also notice that all the languages discussed here look essentially similar. Apart from syntactical differences and variations in the degree to which features such as dynamic binding are available, most OOPLs are conceptually, quite alike.

Among LISP and C, C is more widespread in its use. It is also more efficient in terms of execution speed and memory requirements and is available on a greater variety of hardware platforms. This chapter has focused on, and compared in detail, two of the most popular of the C based OOPLs, Objective-C and C++. Other languages such as Smalltalk and Eiffel have only been briefly discussed. A variety of sources of information on these languages are available to the interested reader.

The reader who is interested in developing a manufacturing related application in an OOPL must consider various factors before choosing a language. Some of the common concerns are as follows (this is *not* an exhaustive list).

1. *Availability of software engineers*: Using languages such as ones based on LISP can limit the availability of programmers. Those that are hired may not be very experienced with the language and may require additional training time.
2. *Number of language vendors*: If the language definition is proprietary to one company then the number of suppliers of compilers and other language related tools tends to be very small (usually only one). This can be risky especially if the manufacturing application will require a large investment and is a long term commitment. Fewer the language related choices available, then the more dependent the developer becomes on a single supplier. This can severely limit portability as well as the number of hardware platforms on which the application will run. The developer's fortunes become too closely tied to those of the language vendor.
3. *Portability*: It is increasingly important that applications run on a wide range of hardware platforms. If the application is written in an OOPL that requires special hardware or the compiler for which is not available on different machines and operating systems, then it could become a liability in the long run.
4. *Efficiency*: Languages that produce code that is more efficient in its use of resources and that runs faster would naturally be preferred. Interpreted languages such as the LISP-based languages would fare poorly here. Hybrid languages would do better.
5. *Integration*: We would like to reuse existing software that was developed without the benefit of an object-oriented methodology and integrate it with newly developed code. An OOPL that can easily be augmented with facilities to interface with other languages will be advantageous.
6. *Object-orientation*: As we have seen in the discussions above, most OOPLs provide varying degrees of the different object-oriented features. The de-

veloper has to decide which ones are important to the application and then select the language that provides those features. For example, artificial intelligence and expert systems applications may require dynamic binding. Encapsulation will be very important when we are trying to integrate new code with older, non object-oriented software. The importance of these feature may force the developer to forego some of the other considerations such as efficiency, or even portability. Perhaps the application will, by its very nature, require only a specific hardware configuration.

In general, different languages are useful for different stages of the development process. As noted earlier in the comparison of Objective-C and C++, the developer may decide to use a language such as Objective-C or Smalltalk for the design and prototyping stages. Then once the class hierarchies and messaging protocols are relatively stable, one could move to a more efficient language such as C++.

This set of guidelines is not exhaustive nor is it meant to be. It is simply a list of issues that are common to most systems. The choice of an object-oriented programming language is one that needs to be made carefully after a detailed analysis of all that is required of the system to be developed.

References

Adiga, S. and Gadre, M. (1990) Object-oriented software modeling of a flexible manufacturing system. *Journal of Intelligent and Robotic Systems*, **3**, 147–165.

Bobrow, D. G. and Stefik, M. (1983) *The LOOPS Manual*, Xerox Corporation.

Boehm, B. W. (1988) A spiral model for software development and enhancement. *IEEE Computer*, May.

Brumbaugh, D. (1990) Object-oriented programming in C. *The C Users Journal*, July.

Cox, B. (1986) *Object-oriented programming: An evolutionary approach*, Addison-Wesley, Reading, MA.

Crabb, D. (1990) ACTOR offers a sophisticated OOP development system. *Infoworld*, October 15.

Digitalk Inc. (1986) *Smalltalk/V: Tutorial and programming handbook*, Los-Angeles, California

Ellis, M. A. and Stroustrup, B. (1990) *The annotated C++ reference manual*. Addison-Wesley, Reading. MA.

Gallagher, J. (1991) Basic concepts II (variations on a theme). *Object-oriented languages, systems and applications*, eds., G. Blair, J. Gallagher, D. Hutchison and D. Shepherd, Halsted Press, UK.

Glassey, C. R. and Adiga, S. (1990) Berkeley Library of Objects for Control and Simulation of Manufacturing (BLOCS/M). *Applications of object-oriented programming*, (eds. L. Pinson and R. Wiener), Addison-Wesley, Reading, MA.

Goldberg, A. and Robson, D. (1983) *Smalltalk-80: The language and its implementation*, Addison-Wesley, Reading, MA.

Guttman, M. (1988) Actor. *Micro/Systems*, July.

166   *Comparing object-oriented programming languages*

Kilian, M. (1990) Trellis: Turning designs into programs. *Communications of the ACM*, **33**(9).

Meyer, B. (1988) *Object-Oriented Software Construction*, Prentice-Hall, Hertfordshire, England.

Micallef, J. (1988) Encapsulation, reusability and extendability in object-oriented programming languages. *Journal of Object-Oriented Programming*, **1**(1).

Pohl, I. (1989) *C++ for C programmers*. Benjamin-Cummings Publishing Co.

Rettig, M. (1987) Using Smalltalk to implement frames. *AI Expert*, **2**(1), 15–18.

Rettig, M., Morgan, T., Jacobs, J. and Wimberly, D. (1989) Object-oriented programming in AI – new choices. *AI Expert*, January, 53–70.

Rumbaugh, J., Blaha, M., Premerlani, W., Eddy, F. and Lorensen, W. (1991) *Object-oriented modeling and design*. Prentice-Hall.

Saunders, J. H. (1989) A survey of object-oriented programming languages. *Journal of Object-Oriented Programming*, **1**(6) 5–11.

Shemer, I. (1987) Systems analysis: A systemic analysis of a conceptual model. *Communications of the ACM*, **30**(6).

Stefik, M. and Bobrow, D. G. (1985) Object-oriented programming: Themes and variations. *The AI Magazine*, Winter, 40–62.

Stepstone Inc. (1989) *The Objective-C Compiler v4.0* and *ICpak 101 v4.0*. Stepstone Corporation, Sandy Hook, CT.

Stroustrup, B. (1987) *The C++ programming language*. Addison-Wesley, Reading, MA.

Stroustrup, B. (1990) On language wars. *Hotline on Object-Oriented Technology*, **1**(3).

Tello, E. R. (1989) *Object-oriented programming for artificial intelligence – A guide to tools and system design*. Addison-Wesley, Reading, MA.

Wiener, R. S. and Pinson, L. J. (1988) *An Introduction to object-oriented programming and C++*, Addison-Wesley, Reading, MA.

Summary: Part Two

S. ADIGA

System architecture

Software objects have the potential to make a similar contribution to the software industry as the packaging design effort in the form of integrated circuits did for the computer hardware industry. Cox (1986) describes reusable, compiled objects as Software-ICs™. The concept of a Software-IC is at too small a level of granularity for organizational level CIM systems. We need a system architecture based on OO ideas in order to take full advantage of the benefits of object-orientation.

Our system architecture for CIM is based on an analysis of applications in modules that are expected to exhibit object-like behavior. The partitioning of applications assumes communication via messages between modules that exist at a higher level than the Software-IC described by Cox. Extending Cox's analogy, we could say that we are seeking circuit-board level reusability as well as chip level reusability through this effort. Integration of applications is enabled conceptually through the use of the same paradigm in system design, and physically by using a common software library to build applications.

Design

Building prototypes and learning from them has been a popular approach to the designing of products in engineering; automobiles have been built in this way for years. OOD does not imply that we should abandon proven methods of testing and implementation, etc. Rather, we have proposed an incremental and iterative approach to building OOS prototypes.

Our approach to prototyping has been one of following a set of good design rules that we have found useful in our practice. However, these rules are exercised iteratively on a set of models: process, data and event. These models provide the context and boundaries of the design process.

We believe that our event-orientation is a useful concept for keeping interaction with the user interface aspects of the program development, and also with user needs. It is quite easy and natural to translate user needs into events that resemble real-life events. This also helps us to relate the inevitable changes in requirements to internal objects as changes take place.

Databases

We share the opinion of Rowe (1987) that stand-alone OODBs do not solve many real-life problems. Generally, one wants to solve manufacturing problems that involve data. Object-oriented databases provide the facilities to store and access objects and class libraries for reuse. These systems must manage storage of complex objects (which may consist of other simpler objects).

Richness of the object model has proven to be a mixed blessing for OODBMS developers. They are faced with new features being required in databases and also modifications to existing techniques of concurrency control, recovery, integrity specifications etc. OODBMS have come a long way from the early prototype designs; with the current pace of research, development and commercialization, they are poised to meet the challenges of the marketplace.

Object-oriented programming languages

The benefits of a good design are translated into gains in overall program effectiveness if there is little mismatch between design philosophy and implementation philosophy. There are great disputes among the OO programming community about the merits of different languages. Though preferring Smalltalk for its OO features, we have picked C-based languages for detailed review on the basis that most real-time applications are, currently, handled more efficiently in a C-based language. In Chapter 7, Milind Gadre has highlighted the pros and cons of C++ and Objective-C, two well-known languages in this area. However, the choice of an implementation language must be made by taking into consideration the operating conditions of the application.

Concluding remarks

The chapters in Part Two have concluded the conceptual discussion on OO technology. Part Three features three industrial applications implemented by two firms who have pioneered early entry into this area, Savoi and Consilium Inc. As early adopters of this technology, they have had to develop their own methods for dealing with software development with objects.

References

Cox, B. J. (1986) Object-oriented programming: An evolutionary approach. Addison-Wesley, Reading, MA.

Rowe, L. (1987) The OBJFADS shared object hierarchy. *Addendum to the Proceedings of Object-oriented Programming Systems, Language and Applications* 75–6, Orlando, Florida.

Bibliography

Ahmad, S., Wong, A., Sriram, D. and Logcher, R. (1991) A comparison of object-oriented database management systems for engineering applications. Research Report R91-12. Intelligent Engineering Systems Laboratory, MIT.

Banerjee, J., Kim, W. and Kim, K. C. (1988) Queries in object-oriented databases. In *Proceedings of Fourth IEEE Conference on Data Engineering*, IEEE Computer Society Press, Los Angeles, CA.

Beeri, C. (1990) A formal approach to object-oriented databases. *Data and Knowledge Engineering*, **5**(4).

Date, C. J. (1990) *Introduction to database systems*. Fifth edition. Addison-Wesley, Reading, MA.

Graefe, G. and Maier, D. (1988) Query optimization in object-oriented database systems: A prospectus. *Advances in object-oriented database systems: Second International Workshop on Object-Oriented Database Systems* (ed. K. R. Dittrich), Springer-Verlag, New York.

Hatvany, J. (1985) Intelligence and cooperation in heterarchic manufacturing systems. *Robotics and Computer Integrated Manufacturing*, **2**(2), 101–104.

Korth, H. F. (1988) Optimization of object-retrieval queries. *Advances in Object-Oriented Database Systems: Second International Workshop on Object-Oriented Database Systems*, (ed. K. R. Dittrich), Springer-Verlag, New York.

Lawson, H. W. (1990) Philosophies for engineering computer-based systems. *IEEE Computer*, December, 52–63.

Low, C. (1988) A Shared, persistent object store. In *Proceedings of ECOOP '88*, (eds. K. Nygaard and S. Gjessing), Springer-Verlag, New York.

Meyer, W., Isenberg, R. and Hubner, M. (1988) Knowledge-based factory supervision – The CIM shell. *International Journal of Computer Integrated Manufacturing*, **1**(1), 31–43.

Pan, J. Y.-C., Tenenbaum, J. M., and Glicksman, J. (1989) A Framework for knowledge-based computer-integrated manufacturing. *IEEE Transactions on Semiconductor Manufacturing*, **2**(2), 33–46.

Shriver, B. and Wegner, P. (eds.) (1987) *Research directions in object-oriented programming*. MIT Press, Cambridge, MA.

Solberg, J. J. (1989) Production planning and scheduling in CIM. In *Information Processing '89* (ed. G. X. Ritter), Elsevier Science Publishers B.V., 919–925.

Tanaka, K., Yoshikawa, M. and Ishihara, K. (1988) Schema virtualization in object-oriented databases. In *Proceedings of IEEE Fourth Conference on Data Engineering*, IEEE Computer Society Press, Los Angeles, CA.

Tsichritzis, D. C. and Nierstrasz, O. M. (1988) Fitting round objects into square databases. In *Proceedings of ECOOP '88* (eds. S. Gjessing and K. Nyhaard), Springer-Verlag, New York.

Valduriez, P., Khoshaifan, S. and Copeland, G. (1986) Implementation techniques of complex objects. In *Proceedings of Twelfth International Conference on Very Large Databases*, (eds. G. Gardarin, W. Chu, S. Ohsuga), Morgan Kaufmann, Los Altos, CA.

Whang, K. Y. (1989) A seamless integration in object-oriented database systems. In *Proceedings of IEEE Conference on Data Engineering*, IEEE Computer Society Press, New York.

Zdonik, S. (1986) Version management in an object-oriented database. In *Proceedings of International Workshop on Advanced Programming Environments*, (eds. T. M. Didriksen, R. Conradi and D. H. Wanrik), Springer-Verlag, New York.

Part Three
Manufacturing Applications

8 FlowStream: An object-oriented plant floor management system

BARRY LOZIER

8.1 FlowStream as a plant floor management system

This chapter describes details of the implementation of FlowStream, a software product of Consilium, Inc., of Mountain View, California. FlowStream is a system used by operators and supervisory staff on a plant floor to help plan, execute and monitor production runs. This product has been designed using the latest software technologies, and in particular utilizes object-oriented programming techniques throughout. This chapter will discuss where and how object-oriented approaches have been used, and the effect they have had on the system's development and delivered functionality.

8.1.1 Application context

Plant-floor management systems are integrated computer applications, targeted to the needs of process industries such as plastics, specialty chemicals, food processing, and textiles. They are designed to fill a niche between corporate planning systems (such as MRP) and automated equipment, providing plant-floor operations staff with timely, coherent data they can use to more accurately control and maintain throughput of manufacturing orders.

This software incorporates the functionality of many existing 'point solution' systems, such as statistical quality control packages, that are usually implemented on standalone platforms. The approach is to offer these features within a common framework – a *software backplane* – which provides unprecedented opportunities for users to correlate incoming data and, thus, to make more intelligent moment-by-moment decisions. Consilium's product strategy for CIM has consistently focused on the critical need for a single data repository for the plant floor.

The system's major areas of functionality are:

1. work-in-process tracking;
2. quality data collection/analysis;
3. plant planning/scheduling – job dispatching;

4. execution control – transaction sequencing; and
5. work instructions display.

To support these features, and to further the cause of CIM implementation, a plant-floor management system (PFMS) must connect with many existing computer systems at different levels within the manufacturing organization. It must interface with corporate-level applications such as MRP, inventory control, and decision support systems, both downloading data to construct its plant-floor model and uploading results of manufacturing runs. It connects to systems in the engineering section for CAD/CAM work instructions and blueprints, or to CAPP (computer-aided process planning) systems for manufacturing plans and product definitions. Often, the interface to these systems passes through a document control system, to ensure appropriate reviews and signoffs.

There are also systems already installed in the plant-floor environment. The process industries have invested heavily in sophisticated real-time control systems. These, in turn, are directly connected to automated equipment such as material handling systems, robotics systems, or simple sensor/actuator combinations. Cell controller technology is also used, particularly in areas of the plant that deal in more discrete work products (like bottles of solution, or barrels of chemical compounds). A PFMS must be effective in integrating these systems into the software backplane, to minimize data entry and to tie in completely and coherently with the existing plant-floor control strategy. Figure 8.1 describes the overall context of systems which affect the plant floor, and where PFMS fits into this hierarchy.

In terms of general computing support, fault tolerance and minimal downtime is an absolute necessity. CIM projects often require scalability, allowing for rapid deployment of a small, prototype implementation; smoothing the way for flexible ramp-up to full production capabilities; and providing for rapid reconfiguration of both hardware and software to meet ongoing needs and to eliminate computing bottlenecks. Return on investment must be demonstrable. In particular, customers will want to make the smallest investment in hardware resources necessary to support the functionality, and will want to make any further investments in small increments.

8.1.2 The FlowStream solution

Consilium has developed a new product, known as FlowStream, to compete in this marketplace. The combination of the flexibility inherent in a distributed architecture and the strong modularity provided by object-orientation meets the challenge of providing a state-of-the-art PFMS. FlowStream will be able to take full advantage of rapidly advancing hardware capabilities well into the 1990s and beyond, especially those offered by workstation-type systems. It provides the means to tailor the software to efficiently use the computing power available within specific network layouts, and to respond quickly to changes within that layout.

Corporate level ('upline')

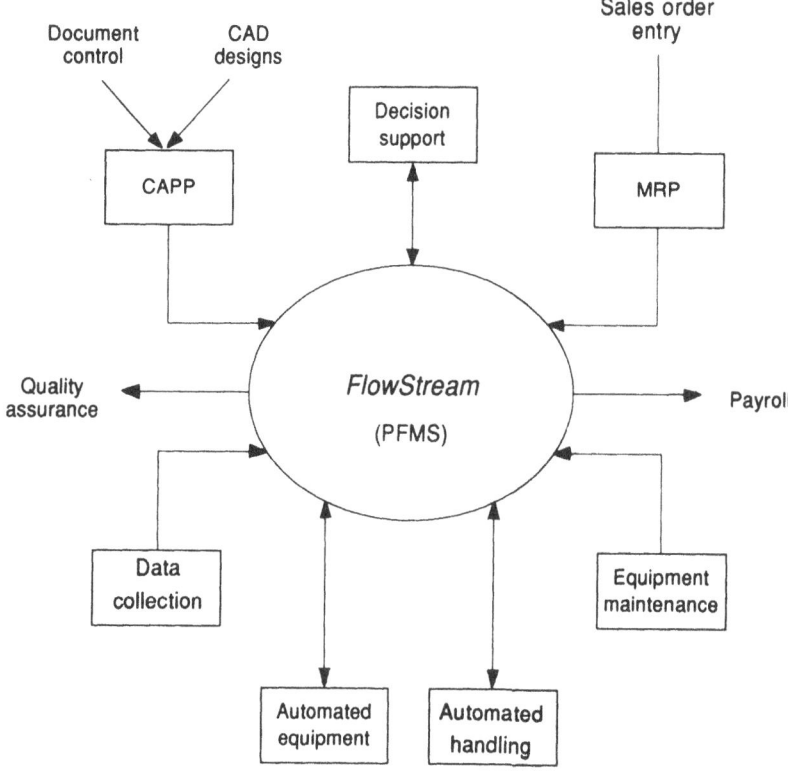

Figure 8.1 *PFMS in the plant-floor system hierarchy.*

Modern software technology has been used wherever possible, to ease system integration efforts, to improve performance and robustness, and to ensure Consilium's development productivity. Industry-standard tools, such as X-Windows, a relational database accessed via SQL, OSI networking, and ANSI C have been used exclusively. These ensure maximum portability across the open systems environment, and safeguard the development investment. FlowStream currently runs under VAX/VMS; a port to UNIX (OSF/1) is planned in future releases.

FlowStream is a multi-process implementation, which uses a client–server architecture to allow for distributing the computing load across networked CPUs. Figure 8.2 is a process-level diagram of a running FlowStream system. Processes respond to events on the user interface, from automated equipment, or from other FlowStream processes. They act as servers for a given functionality – which may,

An object-oriented plant floor management system

in turn, incorporate other lower-level functionality provided by different processes. In general, the idea is to provide a very flexible way of configuring
FlowStream to run on existing computing resources on the Ethernet network, and
to allow rapid changes in configuration to match changes in customer require-

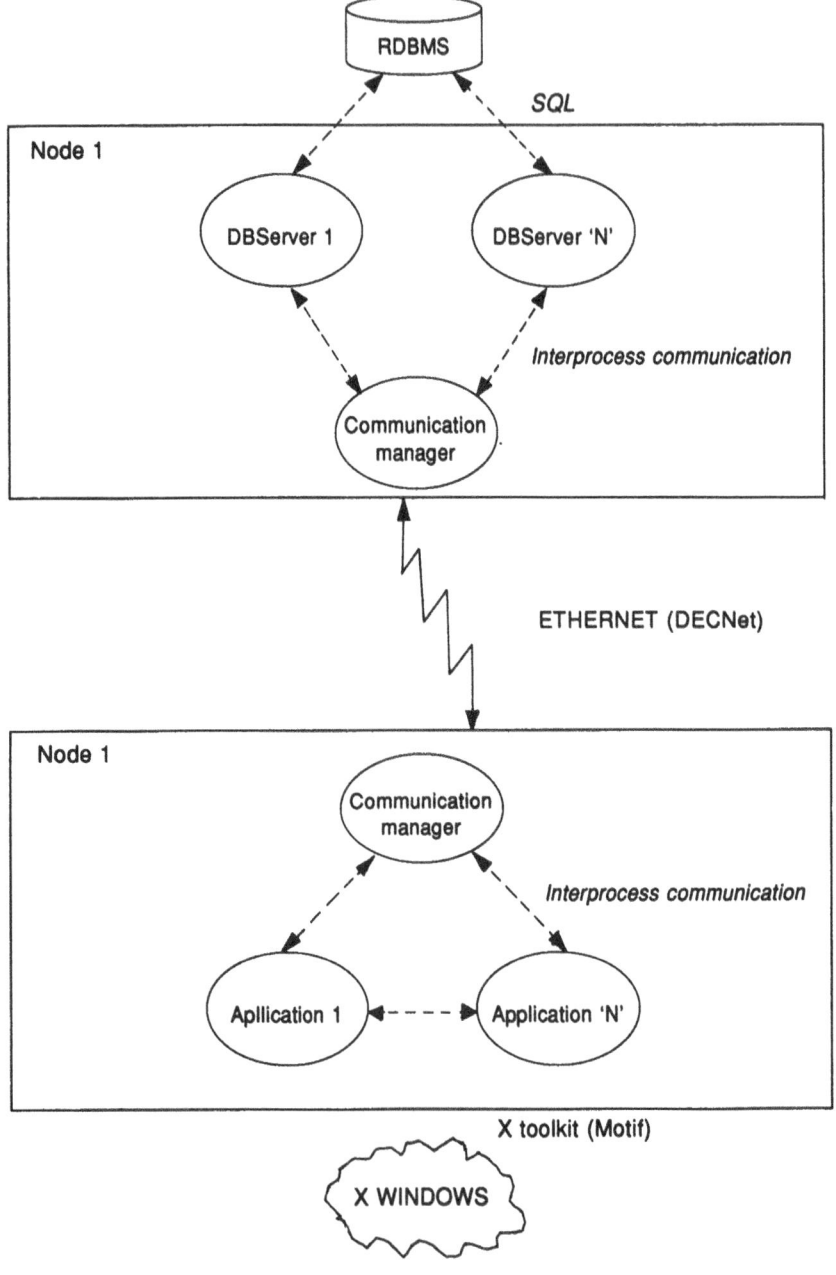

Figure 8.2 *FlowStream processes.*

ments or available computing power. Client–server process configuration is controlled via the FlowStream database, and a given node in the network can be reconfigured without affecting other nodes.

8.2 Rationale for object-oriented development

The FlowStream development project has been active since mid-1988. Very early in the project's life cycle, it was decided that object-orientation should be a primary design and implementation philosophy. This was driven mainly by Consilium's impressions of the target market, and the consequent demands on the development staff. These were formulated many months in advance of beginning the project. The project management staff perceived that whatever development environment was chosen, it had to be able to address three critical requirements – *reusability, portability, and modularity.* Many factors were involved in establishing these as the foremost objectives.

Consilium's markets have traditionally been vertical segments of the manufacturing arena, such as semiconductors, aerospace, and pharmaceuticals. Flow-Stream, while aimed at a different sector altogether from the company's first product, will most likely follow a similar marketing strategy. It has proven very difficult to provide an 'off-the-shelf' plant-floor management system that meets the requirements of any two of these segments; nearly every customer needs some degree of customization – and ultimately product specialization – to build a successful CIM implementation.

For the development staff, this dictated that code reusability is critical to the long-term viability of FlowStream. It was deemed worth the investment to develop a flexible 'toolkit' from which customized installations could be built. The combination of this need for specialization, growing competition within the plant floor system market, and the relatively small size of the company and the development staff, makes the ability to significantly leverage each engineer's contribution a paramount concern. Since the project sets goals for both the application functionality and the technology subsystems that support it, Consilium needs some way of reducing the complexity of those foundation layers for FlowStream application programmers. It was hoped a short term investment in tool-building would lead in the long term to high levels of engineering productivity and the ability to respond rapidly to different market segments.

Just as the need for application sophistication and flexibility drives the development environment, so too do changes in available hardware platforms. Manufacturing executives are willing to make very significant outlays to purchase computer systems which improve their operational effectiveness. By the same token, they must continually justify the existence of these systems on the basis of return on investment. As such, they are very interested in recent developments in workstation hardware offerings, and their continually diminishing 'dollar per MIP' ratios. They see these platforms as providing both an incremental growth capability, and an inexpensive way to deploy ever-greater levels of computational power to address their operational problems.

For Consilium, this meant that FlowStream has to be able to take advantage of these new technologies as soon as possible after they become commercially available. The potential cost of a FlowStream installation drops in direct proportion to the cost of an adequately configured workstation CPU. Thus, FlowStream has to be independent of specific implementations of underlying tools like operating systems, database management systems, and network communication protocols. 'Open systems standards' was the catchphrase.

Code modularity is where these two concepts come together. Since FlowStream would be a set of new integrated applications built on a new technology infrastructure, it was important to isolate these layers from each other from the outset. It was perceived that these two ends of the overall system, while obviously sharing a common overall goal, would grow independently of each other – and in fits and starts. Breakthroughs in technology tools could conceivably come after significant amounts of application code had been produced.

Also, the system will clearly grow to a very large scale in a two-to-three year timeframe. Many of the FlowStream development managers had worked on other major projects in the past, and were aware of the inevitable integration headaches that resulted from such situations – particularly where clearly delineated software interfaces such as protocol stacks were not in place and/or were not closely monitored. Customer requirements brought this issue into even sharper focus. The marketing group had established early on that individual customer site proposals and specifications might in fact dictate exactly which base technologies were to be used. Whereas this notion would not be encouraged, it could play the deciding role in whether a major customer committed to FlowStream.

These prerequisites for the development environment seemed to point naturally in the direction of object-orientation. Most OO approaches designate tool building as the primary mode of implementation, so reusability would be addressed nicely. The choice of C++ as a development language drew upon C's legendary portability across operating systems. Its basis in C also made integration of open systems standards like X Windows, SQL, and OSI a straightforward task. Finally, OO's reliance on *encapsulation* – that is, insistence upon usage of public interfaces as the only legitimate (and, in the case of C++ compilers, the only allowed) way to call upon lower layers of software – enforced strong modularity almost by default. Clearly, this new technology, provided the tools could deliver what the theory promised, was the basis upon which to build.

Once the overall design paradigm was established, the task of determining how to apply C++ to FlowStream's particular problems began. The sections below describe in detail the overall FlowStream object system in its current form as of the time of this writing. Attention will be drawn back to this expository section, since many of the engineering design decisions emerged directly from the original FlowStream product concept and/or the needs of plant floor management customers as perceived by Consilium marketing and development staff.

8.3 The FlowStream object architecture

This product, from its inception, set out to accomplish ambitious objectives in both its application functionality and its technology base, both of which would have to be supported by internally developed code. FlowStream is therefore comprised of two basic components, the application layer and the foundation layer. It quickly became evident, as the first architectural designs were drawn up, that an object-oriented approach could be widely applied in both major areas. The sections below detail the various major classes of objects developed to address conceptual requirements within the two layers. Figure 8.3 illustrates how these objects interact within a typical FlowStream module.

8.3.1 The foundation layer

The foundation layer consists of tools which serve as vehicles for access to the various technologies available on the runtime computing platform. These include the operating system and related services, the means of interprocess communication and network access, the relational database system, and the mechanics of the user interface.

Operating system (OS) processes are modeled as '*Process objects*'. This allows for modularization of all code related to creating, maintaining, and destroying a process on the specific target OS. The particular modality for interprocess communication and network access is also implemented here, since these two notions are often closely connected with the choice of operating system. The current incarnation of this object class supports VMS under DECNet. Note that this tight OS–IPC coupling may not always be the case; but C++ polymorphism provides a way of building a UNIX–DECNet object or a VMS–TCP/IP object at the lowest levels of code, and using them interchangeably as generic process objects in the upper layers. Polymorphism and its benefits will be discussed in more detail below.

The process object works with a class of objects known as '*Events*'. These are essentially message buffers that contain protocol 'headers' and other data to allow for routing to other processes. The term event (as opposed to 'message') is used to denote the fact that processes are event-driven – their lives are spent responding to events, often by generating outgoing events of their own. On the sending side, processes create and send events as needed to implement the client–server architecture. It is important to note, however, that most processes in such an environment act as both clients and servers during their lifetime. Processes prepare to serve requests from other processes by registering callback functions, which constitute the 'reaction' of that process to the receipt of a given type of event. Figure 8.4 illustrates the flow of events through a FlowStream process.

Necessary OS support boils down to multi-process support: the ability to create new processes, and to allow them to communicate. By centralizing this functionality in a single object, porting FlowStream code to other operating systems and/or network protocols is greatly assisted. Adding the concept of events and event-driven behavior, which seemed an obvious derivation of the basic process object functionality, essentially completed the support for a client–server architecture.

C++ objects have also been utilized at the user interface (UI) level, with a major effort devoted to encapsulating common interaction themes for reuse throughout the various applications. These basic '*UI component objects*' can then be subclassed into X-Windows type, barcode-type, terminal-type, etc. These may

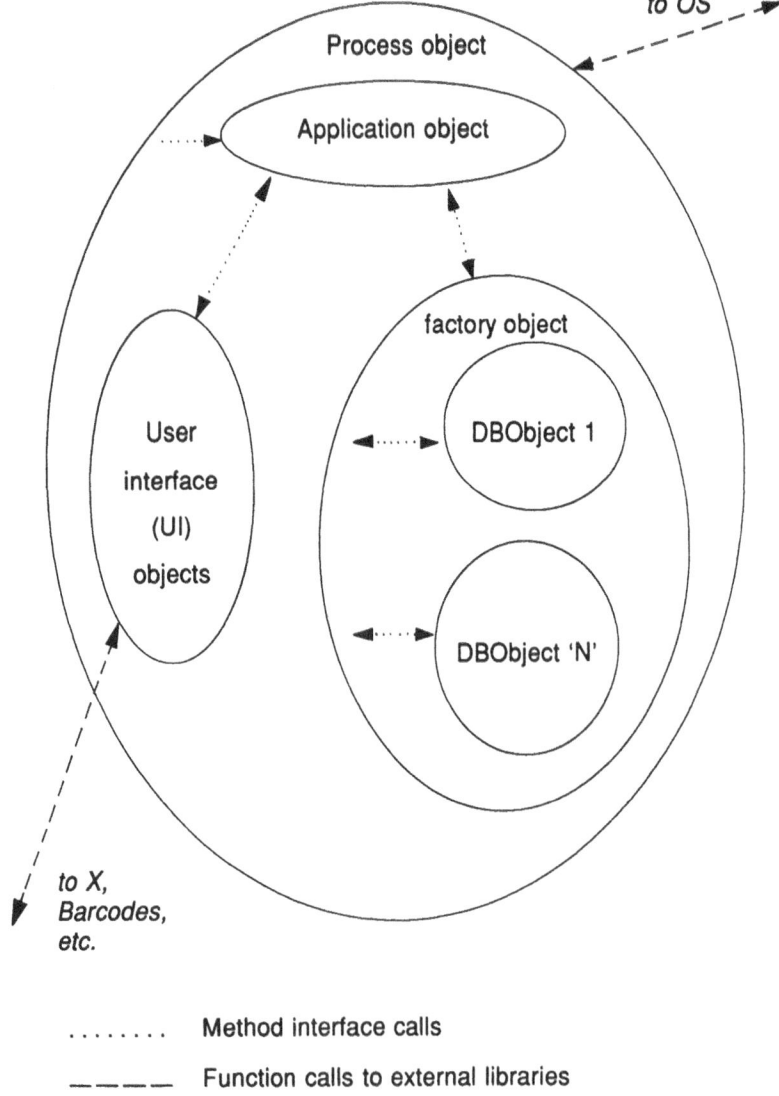

.... Method interface calls

– – – – Function calls to external libraries

Figure 8.3 *Object interactions within a FlowStream application module.*

actually implement the given theme in very different ways, but provide an identical interface to applications. Thus, these applications can be written with only minimal dependencies on the target UI display device.

FlowStream's primary interface will be X-Windows. The Athena project at MIT, which originated this package, defined X-Windows in an object-oriented fashion, but implemented it in C. The hierarchy of X 'widget' objects can be extended and maintained in a similar (but more complex) fashion to C++ object classes. However, the FlowStream development team decided to use C++ objects

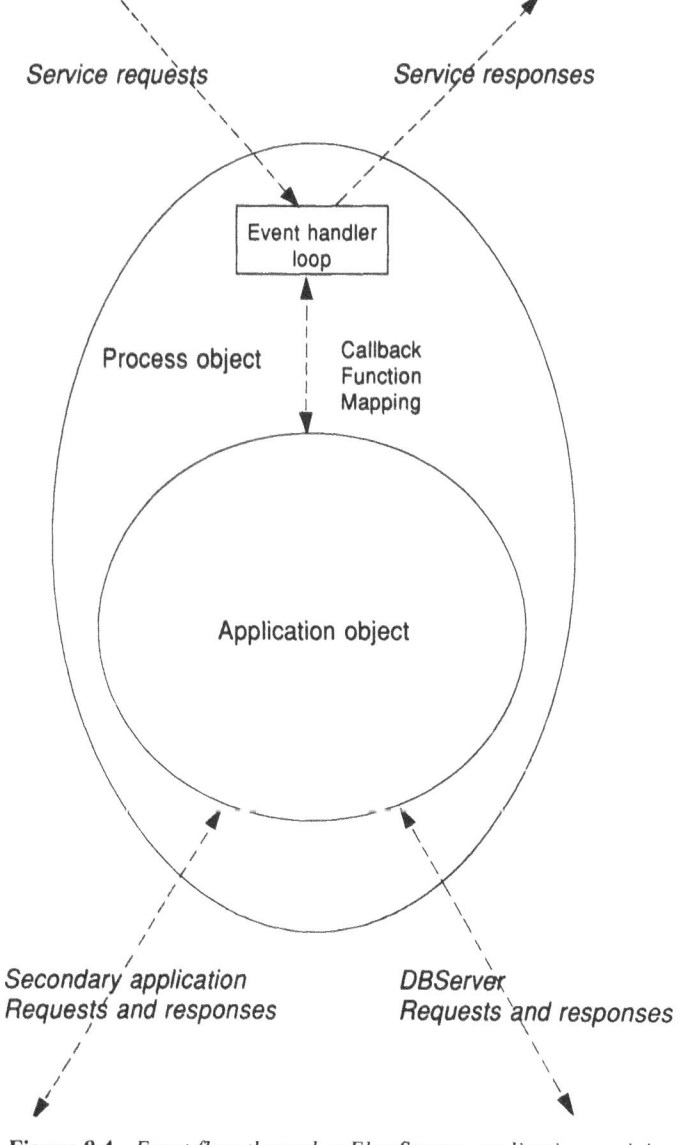

Figure 8.4 *Event flow through a FlowStream application module.*

directly to compose user interface tools. It was very desirable to provide, at minimum, a C++ interface to UI components, since that was the environment in which the remainder of the applications would be coded. Since it was clear that FlowStream would have to support multiple types of display devices very early in its lifecycle, it was understood that X could not act as the generic API (application programming interface) to support user interaction.

An example of a meaningful theme would be a 'text-field' object: a text entry field, preceded by a text label description. Some combination of X-Windows widgets are used to implement this functionality under X, while the barcode equivalent issues whatever commands are appropriate for some specific target barcode reader device.

This is not to say that some applications do not require revision to support different target interfaces – only that the UI component tools make it possible to write applications which will span these interfaces without modification. Creation and use of these UI objects has been straightforward, and confers the benefit of consistency of interaction methodology across all FlowStream application modules. What this means for FlowStream users will be elaborated in a later section.

Tools for database interactions are also provided in the foundation layer. Database tuples, key structures, lists of tuples matching a given selection criteria, and transactions (atomic blocks of work against the DB) are all modeled as C++ object classes. As such, application-level code performs all retrieval and modification of persistent (stored) data via these vehicles. This serves to insulate them from database-specific implementation details like SQL syntax and embedded language support; multi-user locking considerations, except in the most basic sense; and the means of efficient navigation through the normalized database schema, whose structure may appear somewhat different from application data structures (more on this below).

These objects are used as both the application interface to the database, and within a specialized DBServer process as the direct linkage to the target RDBMS. Since each database relation is an independent data structure expressed in the database definition, separate subclasses of a common base class (*DBObject*) must be written to support each individual relation. However, the combination of C++'s generic class capability (essentially a template, which is expanded for a specific instance using the compiler's preprocessor) and an internally developed parser allow the code for these objects to be generated directly from the data dictionary, with no programmer intervention necessary.

Thus, a powerful, versatile foundation of operating system, network, user interface and database support is established, upon which application functionality can be layered. This technology has been designed to address the three objectives of reusability, portability and modularity. Indeed, these notions are much more important at this level of software, since the much larger application layer will express its implementation in terms of foundation-level objects, whereas the foundation objects depend directly on platform-specific capabilities and interfaces.

8.3.2 The application layer

This software layer consists of factory objects, which model basic plant-floor concepts and entities; and application objects, which model particular transactions or units of work on the plant floor. Application objects combine various factory object interactions to produce particular product functionalities. Figure 8.5 superimposes this interaction onto Figure 8.4 (the process–event flow) to give a more complete picture of how a FlowStream process is internally structured.

Factory objects

The first, most important set of objects encountered at the application level are the factory objects. These are truly the building blocks of all FlowStream applications, since they model real-world entities on the manufacturing plant floor. These include physical entities, such as laborers, pieces of equipment, and inventory storage locations; and logical entities, such as formulae, process instructions, and quality assurance testplans.

This software layer is the consumer of the foundation's database objects. Factory objects use these objects to instantiate themselves from secondary storage (i.e. the RDBMS), and to update their persistent representations as necessary to reflect application-level actions. The applications themselves are written strictly in the form of factory object behaviors and interactions. A given factory object translates these into database interactions as appropriate.

A critical aspect in the design of factory objects is that their data structures are hierarchical. Another way of saying this is that factory objects exist only in the dynamic memory of some application process; as such, they can take full advantage of object-oriented data structuring capabilities like inheritance hierarchies, as well as utilizing C memory pointers as instance data. This differs from the constraints placed on the definition of database objects, which are used primarily as vehicles to transport data back and forth between the DBServer process and the application. To meet this 'shipability' requirement, database objects must be flat data structures – they cannot contain references to pointers, and (by virtue of their modeling a single tuple in a single database relation) they are generally not broken into hierarchical components.

The disparity between factory object and database object data representation also centers around the notion of *consumers*. Consumers of the object-oriented factory object model include FlowStream application programmers, and customer staff who are working at source-code level to build custom FlowStream-based applications. These individuals desire a fully flexible object environment, typically in dynamic memory only, where they can utilize the full capabilities of the C++ language. A different constituency altogether is the consumers of the relational database model, which includes system integration engineers at customer sites (recall that the FlowStream database will often be used as a software backplane for integration of other CIM components), and end-users accessing the database to produce periodic reports or for problem investigation and analysis. These users want a coherent, straightforward relational model, which places data into as few locations as possible and minimizes the use of mechanical constructs such as contrived key data.

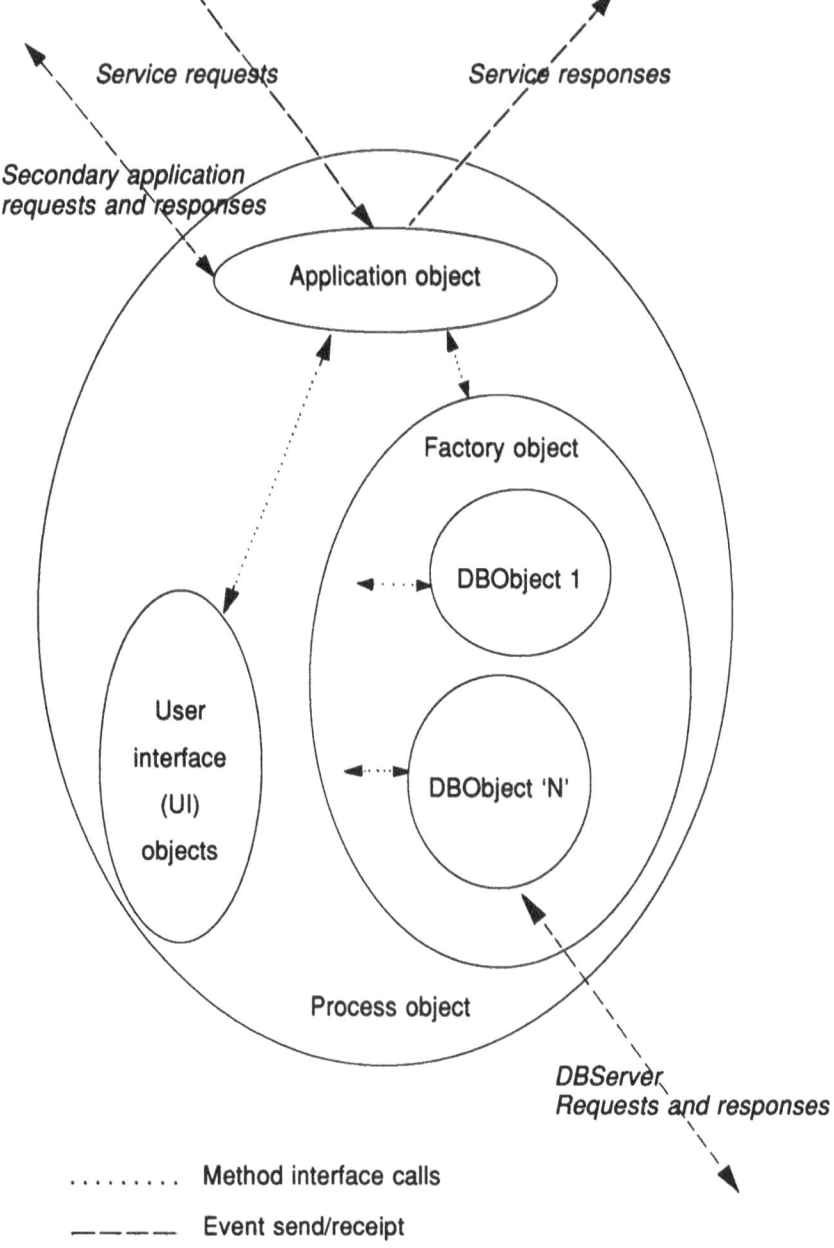

Figure 8.5 *Event flow and object interactions within a FlowStream application module.*

FlowStream development experience over the last three years has indicated that these disparate needs are best addressed independently. More specifically, attempting to model the inherently hierarchical and free-form structure of factory objects directly in the relational schema often leads to conceptual problems at that level. For example, take the case of defining a 'labor factory object' as a type of 'physical resource'. There are many possibilities for schema definition here, but virtually all of them have deleterious consequences for users that access the database directly. One set of possibilities leads, through database normalization rules, to many more relation definitions – which equates to more complex queries and/or view definitions. Another leads to the definition of an 'entity type' field in some relation, which also makes queries more complex and, perhaps more importantly, inefficient. A third leads to the use of contrived keys, in place of names that are recognizable in the real-world environment.

In general, it is difficult to model the notion of hierarchy that C++ supports inherently in the relational database, while at the same time providing an intelligible database schema for users that are accustomed to working at that level directly via report generators, query languages, and other data access tools. Usage of more sophisticated C++ structural facilities, particularly multiple inheritance, will only exacerbate this problem. For these reasons, it was decided to address these requirements separately, and to incorporate into factory objects the means of converting between the two data representations. At the time of writing, significant problems have not been encountered as a result of this direction.

Once this decision was made, it became possible to design factory objects which utilize the complete range of C++ abstraction features. Many different inheritance hierarchies can be deduced within the problem space of the applications:

1. *Equipment* as a type of *resource*;
2. *Bill-of-material* as a type of *formula*;
3. *Workcell* as a type of *facility*;
4. *Floor stock* as a type of *material*;
5. *Process instruction* as a type of *graphically-displayable document*.

These can be implemented directly in the form of parent–child relationships between the factory object classes. Doing so allows applications to be written using *only* the abstractions, using polymorphism. For example, imagine a resource allocation module that can potentially deal with any present or future instance of a resource factory object; a graphical display driver which can handle any type of graphically-displayable factory object, such as statistical process control charts, CAD blueprints, or even FlowStream user documentation; or a flexible notion of 'facility' which allows for complex nested definitions of data and transaction security clearances.

Since factory objects model the real-world entities acted upon by the applications, their usage results in consistency in the treatment of fundamental application concepts. Over the course of Consilium's ten-year history of developing plant-floor applications, it has become painfully clear that the size and complexity of these systems can rapidly lead to differing interpretations of these concepts. The issue is not so much data representation (which is addressable through specific design techniques) as much as behavioral assumptions – which, in functional programming approaches, are hardcoded into the logic of particular transaction modules. The behavior of a particular real-world entity ends up being modeled by small portions of many different applications, and surprising disparities in underlying assumptions often emerge. This can make successful maintenance of these systems difficult at best, since correcting perceived problems in one area often violates the premises of another. Modeling these entities as objects, and encapsulating their behaviors here, has positively influenced the clarity of design concepts, and has provided the ability to explicitly change certain assumptions with assurance that all applications will reflect these changes.

This idea has been expressed in the software engineering literature as the 'principle of minimal commitment' (Miller, 1990). This essentially states that upper layers of software should depend as little as possible on the specific implementation of lower layers. Factory objects are a powerful instance of this principle in action. FlowStream application developers are enjoined to use only the public interfaces of factory objects to build the logic of their specific modules – and C++ provides versatile ways to control this at compile-time. Further, it is strongly encouraged that application behavior be implemented directly within factory objects whenever possible, the best scenario being new behavior built through reuse and/or recombination of existing behaviors.

Beyond the obvious coding benefits, factory objects also serve as the basis for conceptual discussions of application functionality. By interpreting the various plant-floor activities as factory object behaviors or interactions, a lexicon of familiar terms and concepts is introduced to FlowStream developers and designers. Even Marketing and Sales staff begin to refer to the object-oriented view. This will have very positive implications in the long-term, as Consilium attempts to address new industry segments or new areas of functionality.

Application objects

Application services are themselves modeled as objects. These are child classes of the process object, and an instance of some specific application object is declared in the main ( ) modules which constitute standalone FlowStream programs. They combine factory objects into meaningful plant-floor activities, and are probably best understood as the classic notion of a 'transaction screen.'

Database locking modes are controlled at this level. Although the factory objects are the actual mode of interface to specific database information, the plant-floor activities are more logical 'units of work' in the database sense. Since this layer is the consumer of factory objects, it is here that the necessary knowledge resides to indicate whether database access will be read-only or read–write, and the proper time to commit or rollback changes made during the latest database transaction. The modality used by the application objects to perform these functions is a database transaction object, which provides a simple application-level interface, and encapsulates the necessary message traffic to the DBServer process.

Application objects are also the consumers of the UI component objects. They instantiate these objects as needed to implement the specified user interface (on a given target device, usually X), and pass data between these objects and the factory objects. The structure of application objects allows them to be subclassed on the basis of which target UI device they support. Thus, the application logic can be common code, which the child classes of that application tie together with the specific UI component objects needed for a given device. This significantly leverages the application programmer's efforts, since FlowStream applications must currently support a minimum of three types of UI devices.

Application objects are the top layer of code in the FlowStream system. They tie together all of the tools available in the lower layers (both application and foundation-level) to build the actual delivered product. They obtain major benefits from the object-oriented approach, and are relatively simple to implement as a result. They are also themselves reusable to some degree, since the event-driven environment allows one application process to call another.

8.4 Assessing the object-oriented approach

Consilium's experience in utilizing the OO paradigm has been generally positive. The original impetus was the desire to obtain higher productivity and quality of developed code. While it may be too early in the project to be unequivocal, this objective has been reached to a large degree, and the methodology promises future cost savings. Other significant advantages have also been confirmed, some of which we discovered only after beginning the OO development process.

By the same token, much has been learned about the implications and costs of OO over the course of this project. While not seriously diminishing the value of the benefits in the long run, these lessons have been painful at times, and have to be understood and addressed before the benefits can truly be achieved. Adoption of an OO philosophy often involves substantial shifts in design thinking, organization, and procedures for an existing development organization, whose impact should not be underestimated.

8.4.1 Advantages

Nearly all of the various players in the software development process at Consilium, both within the company and at customer sites, derive major benefits from the use of OO to build the FlowStream product. These are generally variations on the efficiency of making changes to the code – the speed with which this can be done, the elegance of one change leading to major modifications across the board, and the relative safety with which new concepts and/or new technologies can be introduced. Users clearly gain by the FlowStream staff's ability to react, but obtain benefits at their own level as well.

Development cost savings

Modeling software in an object-oriented fashion makes enforcement of the 'minimal commitment' principle almost automatic. A high degree of code modularity is the result, which in turn confers many cost savings to the development team. Functionality can be easily extended within lower layers of software, without requiring changes to existing code at the upper layers. As a result, the product as a whole can adapt more easily to changing development and/or runtime environments. In general, any necessary rework of a particular area of the code can be done more quickly, and the potential risk of invalidating other areas can be minimized.

Given the competitive environment of the market Consilium serves, and the nature of the operations FlowStream will perform for manufacturing customers, upgrades to new hardware platforms and rapid adjustment of functionality to meet changing requirements will be a fact of life for this product. The development team must expect to incur these costs in any event and, therefore, any potential cost savings will have immediate bottom-line impact. This positive effect can (and has, in Consilium's case) more than offset the cost of getting developers over the learning curve in the new methodology, and the risk factors that accompany this curve.

In the case of FlowStream, the set of factory objects probably illustrates these advantages most clearly. Recall that factory objects embody real-world entities on the plant floor, and their related application behaviors. Reusability of basic application concepts – coding them in a single, well-defined place – has become an inherent implementation philosophy. Since these behaviors are encapsulated inside a single object, changes or enhancements to them can also be performed there, with assurance that these new definitions will affect any and all appropriate areas of application code. (This is less true in the case of modifications to data structure, but the factory object model still allows identification of areas that need maintenance in a single compile and link of the product.)

The clarity of the factory object model reduces confusion, and makes it possible for teams developing different applications to avoid working at cross-purposes. Minimal commitment allows the main application objects to be as flexible as possible in allowing changes to the implementation of basic application concepts; with the exception of data structure modifications, changes to user interface code are generally not required. Perhaps most importantly, bugs noted in the definition of these behaviors, which potentially affect many different applications, can be corrected efficiently and with a high confidence level that all affected code will be repaired.

The concept of polymorphism – working with an object as an instance of its ancestors – introduces a second level of modularity, not readily obtainable through traditional programming techniques. It leads to a development environment where data structures can often be as modular as algorithms; that is, where upper layers of code can take advantages of the similarities between certain structures, instead of having to concentrate on and adjust for each individual instance. New possibilities for generalization emerge. For example, the ability to deal with generic resources makes development of resource tracking capabilities much more straightforward than a model which is forced to view this as a combination of equipment tracking, labor tracking, tool tracking, etc. Modules can be specific (like an equipment object editor) or generic (like ChangeResource-State) as required. Single applications can be stretched to fit many different objects at once. Perhaps most dramatic, new types of resource objects can be introduced with no changes to existing applications that use the generic parent resource.

Reduced development cycle time

This investment in code reusability and modularity will pay off in the longer-term as reduced cycle time for new and enhanced applications. This is a distinct competitive advantage in any marketplace, but particularly in a situation such as that faced by FlowStream, where customization is the norm and large investments depend on technology and functionality enhancements.

Encapsulation and a strictly object-oriented approach within the foundation layer allows for faster and less chaotic adaptation to new hardware platforms and operating systems. For example, the process object is used throughout Flow-Stream to provide OS-related services. In porting to a new OS, the implementation of process may change fundamentally, but the interface to applications can remain unaffected. The same advantage is obtained when moving to new base technologies for the user interface, database, or communication layers.

Similarly, the use of factory objects to encapsulate basic concepts of the application model makes FlowStream more responsive to specific market segments that may require modifications or extensions to this model to accurately portray their manufacturing environment. Development of new, customer-specific application modules can also occur more quickly and easily, since so many of the existing components can be reused. This speed and flexibility in responding to specific customer requirements is a boon to Consilium's vertical marketing strategy, one which could often turn the tide in the competition for a major implementation contract.

This high level of modularity also bodes well for Consilium as a relatively small company, since it leads to more effective utilization of the existing engineering staff. Junior engineers can be highly productive members of the team almost immediately, since their first assignments can involve projects which use the existing tools to build specific applications, before actually engaging in tool-building itself. The variety of tools necessary to support the ongoing effort makes for a 'rich' development environment, where there are more opportunities for specialization and, conversely, for cross-training. Engineers can gradually learn more and more of the underlying layers – some starting at the foundation level (based on some technology expertise) and working upward, others starting at the application level and moving toward the technology base.

Mission-critical functionalities can move forward under the aegis of established experts, while the contributions of less adept engineers can still be significant and positive. Indeed, this environment has prompted the FlowStream management team to establish a matrix approach to staff allocation to specific application projects. Project teams are created, mixing application-level expertise with specialists from the foundation-level group. This concept minimizes the external dependencies for a given application project, and allows quality work to proceed at the fastest pace possible.

User-level benefits

FlowStream users benefit as much as the development team does from the usage of object-oriented techniques. The most immediate impact is felt at the user interface level. Usage of UI objects to package certain types of interaction themes has resulted in broad consistency in the look and feel of FlowStream across the various applications. This clarity and coherence of presentation makes applications easier to learn. Users can immediately apply their understanding of the particular interaction themes of one application when they see the same theme on a second application. This empowerment of the end-user definitely improves the acceptance of the overall product by individual employees on the plant floor. It allows these users to shift their focus more rapidly and confidently from the mechanical details of interaction onto the application concepts and functionalities.

The user interface level of FlowStream is also more easily maintained. This translates directly to a benefit for users, since the product can incorporate new and better UI technologies – and make them available for use at customer sites – more quickly. Any hardware or system software dependencies are embedded in the foundation code, where they can be changed in one place to affect the entire application suite. In cases where major changes in interaction style occur, based on a change in the lowest-level UI tools being used, this ability to locate and modify a single object which, in turn, changes all places where this interaction style is used throughout the applications can save literally months of coding effort.

For example, FlowStream is now in the process of being ported from DEC-Windows 3.0 to OSF/Motif. This project, which will change significantly the look and feel of approximately 300 000 lines of application code, has been conservatively estimated as requiring 12 man-weeks of effort. This compares to the current level of investment of approximately 100 man-years in the total product.

8.4.2 Caveats

This glowing review of the impact of object-orientation has not come without some headaches. These can be generally categorized as paradigm shift, and its effect on a software team that is mostly rooted in structured, functional design and implementation techniques. The level of support of these new concepts – technical, organizational, *emotional* – is critical to the success of such an endeavor. This support has varied in degree over the course of the FlowStream project, and this variance has brought certain issues to light. These issues are potential 'show-stoppers'. They have not yet had this effect on the FlowStream project, but must be addressed continuously in order to achieve the aforementioned advantages.

Moving from a functional world view to an object-oriented one can take a long time, particularly in an existing development organization with an existing (i.e. structured) development methodology. To be object-oriented involves much more than commencing to write new code in the form of objects. The true advantages of the technology come with the usage of advanced concepts like inheritance and polymorphism. Where these can be successfully applied must be discovered during the design process; and the learning curve for the chief designers of a given product to change their thought processes can be considerably longer than the time it takes for developers to learn an OO programming language.

In the case of C++, the available development tools are limited, and are still in the midst of substantial evolution. There are relatively few vendors that offer C++ tools, although the number has grown significantly within the last year. Integrated environments, such as those available for C where the compiler, debugger, and code profiler/browser are tightly coupled, are still largely vaporware.

The FlowStream management team is still firmly committed to C++ as the primary language of development. However, that decision was based on the potential inherent in the language as an OO platform, not on the maturity of the language definition or the availability of development support products. This life on the 'bleeding edge' can make establishment of stable coding environments difficult, and periodic retraining and rework a necessity. This inevitably leads to some fear and loathing on the part of upper-level management.

Since iterative development is the norm in OO environments, substantial lead time must be allotted for the necessary tools to be developed – and redeveloped. Code reuse becomes the first major objective of the programming team. This cannot be achieved without investment in design work and prototyping. A traditional top-down view does not always surface opportunities for reuse; this, in turn, leads to iterations on various prototypes and design models as the overall concepts come into focus.

Evolution of one layer of the system through these iterations often creates the need to modify downstream layers to accommodate. Indeed, in the earliest stages of FlowStream development, this evolution was usually applied to the interface of a particular object. Recoding (or, more desirably, redesign) of other software layers was often necessary to adjust to the new 'look and feel' of the object. One very important lesson learned here was to avoid delay of these downline modifications – which can be very tempting when the project is under time pressure.

This delay can destroy whatever clarity the model may possess. A recursive, spiraling effect of confusion and miscommunication (leading to more rework, and thus more downline impact) is the result.

An incremental approach seems to be best. A great deal of time should be spent prototyping and establishing the model and the basic tools. Pushing too hard too soon can lead to a real mess, from which it may be more difficult to escape than in a functional methodology. Object class functionality can rapidly become unclear; to meet the demands of time pressure, these will be designed and implemented many different places in many different ways. Poorly designed object-oriented code can be harder to correct than functional code, due to the complexity of interaction between objects. In this scenario, realization of the primary benefits, such as modularity and advanced features like polymorphism, is virtually impossible (or sometimes even disastrous).

The essential message here is to allow for a relatively long ramp-up time before pushing for production-level objectives. It will take time to describe the application problem space in terms of objects, and to develop the necessary objects to support applications at the technology level. Advances in one area will result in major rethinking of concepts in another. Perhaps most significantly, changing the frame of mind of code designers so that advanced OO concepts are leveraged effectively cannot be done through training alone. Coding experience, either direct or vicarious, is often necessary to gain facility with these new ideas.

Finally, object orientation will eventually tend to pervade the design thinking of a development team. This can be both a boon and a detriment. As mentioned throughout this paper, the technology layers of a software system can and should benefit as much as the application layers from an OO approach. However, it is here that existing (probably functional) code is often encountered that needs to be merged with the OO design. Sometimes this is unavoidable, since redevelopment of these modules may not be cost-justified, or supportable by the existing programming staff in the short term. The easiest way to integrate this code is to develop object 'wrappers' that provide an OO interface to a particular module or functionality. These wrapper objects can be difficult to integrate into the overall object model, particularly one that is making use of advanced language features.

The decision to utilize object orientation must be a part of an overall product strategy, that considers lead time and quality objectives as paramount. The OO technology cannot generally be 'sold' to end-users. The substantial upfront investment necessary must be shown to translate to results the customer can see, since the customer is typically unaware and uninterested in what methodology the development team has used to achieve them.

Summary

The FlowStream product, and Consilium as a company, has benefited greatly from the usage of object-oriented techniques in its development. It has resulted in a solid, coherent design and code foundation for the creation and enhancement of application functionality. It has conferred distinct competitive advantages for Consilium, and (consequently) for the users of FlowStream. It has led to an implementation which should flexibly evolve and grow continually through the coming years. This will lengthen the product lifecycle considerably and, thereby, ensure a more-than-adequate return on the development investment.

Whereas they have carried a significant learning curve, and have necessitated a certain volatility in the coding environment, the resulting changes in organization and thought process have left the team stronger and more capable of dealing with change. This may be the biggest benefit of all, since it will allow the team to adapt to the inexorable increases in reaction speed and quality that Consilium customers will come to expect in the years ahead.

Reference

Miller, Ross M. (1990) *Computer-aided financial analysis*. Addison-Wesley, Reading, MA.

9 OOPS in real-time control applications

DAVID WILCZYNSKI and DAVID K. WALLACE
Savoir, 300 Manhattan Beach Blvd., Manhattan Beach, CA 90266, USA
Savoir, 5025 Venture Drive, Ann Arbor, MI 48108, USA

9.1 Introduction

Two real-time control applications are featured in this chapter. In the first, a kitting cell, object-oriented techniques were used to design and implement its control system. In the second, a glass line, similar techniques were used to model it in order to improve its performance and upgrade its code.

Both are typical real-time control applications. Each is implemented using a variety of controllers and sensors. The ubiquitous programmable logic controller (PLC) is in both, as are controllers native to specialized devices (robots, CNC machines, etc.). Where required, the controllers are linked using wires and proprietary networks. People appear where automation is impractical. Programming is done in a disparate set of device-specific languages by the device vendors.

Despite their architectural similarities, the two applications would benefit differently from the object-oriented methodology. In the kitting cell, it was the way to build a rapid prototype in which to study the mechanical interactions and overall throughput of the system. In the glass line effort, the object-oriented analysis would lead to an after-the-fact understanding of the existing system and suggest a variety of incremental improvements.

In both applications the object-oriented methodology was key, not only in fulfilling the technical requirements, but in giving the mechanical engineers a formalism that easily captured the expression of their needs and constraints. Real-time manufacturing control systems yield naturally to these object-oriented methods.

9.2 The kitting cell

9.2.1 Specification

The kitting cell assembles kits full of parts for use by other assembly cells. A schematic of the Kitting Cell is shown in Figure 9.1. Each kit holds anywhere

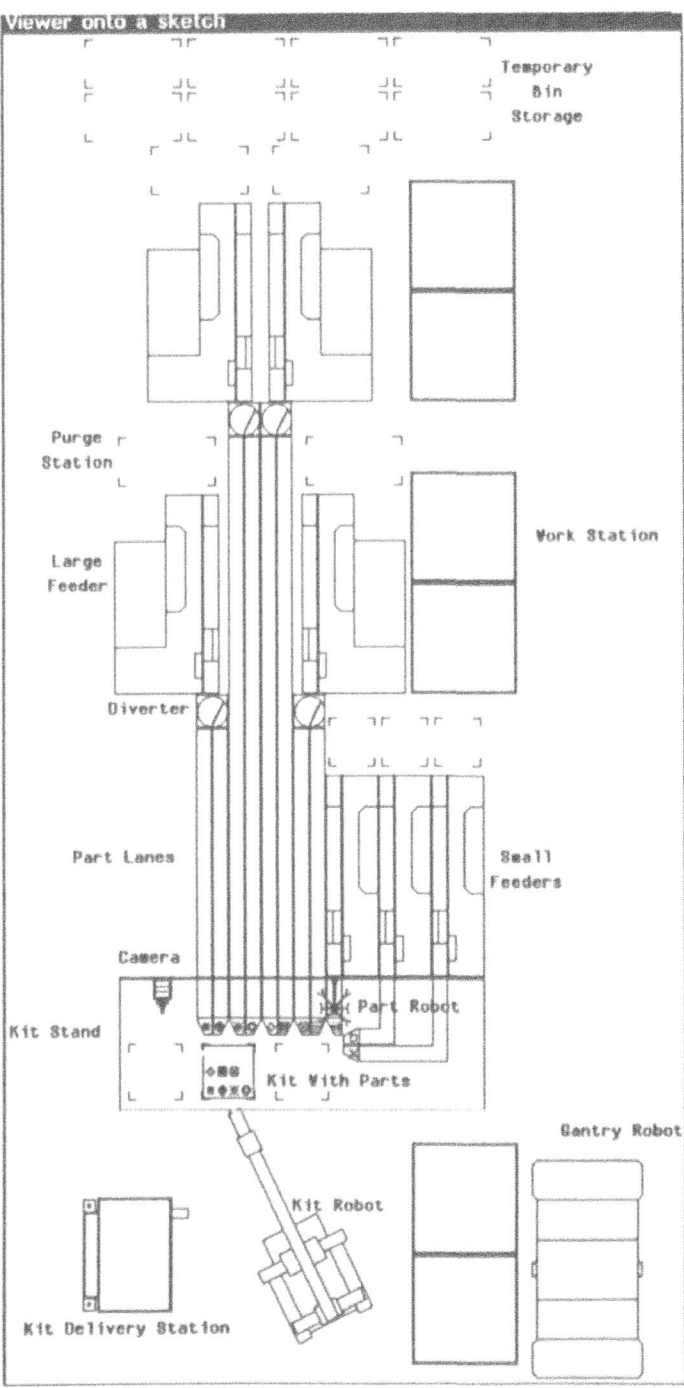

Figure 9.1 *The kitting cell.*

from nine to eleven various parts depending on the configuration in effect. The basic operation of the cell has three divisions: kit management, feeder management, and lane/nest management.

1. *Kit management*: Kits are brought to the kit delivery station by a conveyor (not shown in the figure). The kit robot picks up the kit and puts it on one of the two furthest-right positions on the kit stand. When the kit is complete, the kit robot moves it to the furthest-left position for a camera inspection. If successful, the kit robot moves the kit back to the kit delivery station where it exits the cell.

2. *Feeder management*: Bins holding a single kind of part are dumped into the feeders by the gantry robot. The gantry puts the empty bin in the purge station behind the feeder. The large feeders put the parts one at a time onto a vibrating panel. A diverter orients the part toward one of two long lanes. From there the part vibrates down the lane. Parts queue up behind one another until they arrive at the nest at the lane's end. The three small feeders each deposit a part onto its corresponding short lane. A single drive then vibrates them into the nests.

   Since the large feeders service two lanes, they must be purged and refilled as a lane empties. The feeder drops its rear gate and empties itself into the purge bin. The gantry robot moves the purge bin to a temporary storage location and refills the bin with the required part. The lanes continue to vibrate parts to the nests during this purge cycle.

3. *Lane/nest management*: Parts vibrate down the lanes into the nests. In the long lanes, the lineup of the parts behind those in the nest creates a stable state. Depending upon the part size, those nests may hold anywhere from one to ten parts. The short lanes only feed a single part to their corresponding nests. A stable state is created by turning off the short lanes' drive.

   An overhead camera takes pictures of the nests in order to determine part quality and coordinate information. If satisfactory, the part robot, sliding along a rail over the lanes, uses one of its six grippers to pick up the part. After picking up a load of parts, it puts them in the kit. Generally, two passes are required across the lanes to complete a kit.

   All eleven nests can be individually dumped in case no good parts can be located by the camera. The long lanes, which are generally always vibrating, must be stopped before a nest is dumped.

The feeder/lane/nest hardware was to be controlled by a single large PLC. The hardware had the following capabilities:

1. *Feeder*: On/off switch, vibration amplitude setting, part low sensor, feed parts switch, part fed counter, lower/raise rear gate switch.
2. *Long lanes as a pair*: On/off switch, vibration amplitude setting, diverter left/right switch.
3. *Short lanes*: On/off switch (one for all three).
4. *Nest*: Up/down switch.

The cameras were driven by a controller with the following capabilities: initialize; load configuration information; shoot camera x.

The kit and part robots each had a controller with a normal set of robotic capabilities. The 'packaging' of their code was yet to be determined. The gantry robot and kit conveyor system were shared resources whose messaging protocol was to be determined.

A cell controller would coordinate all these devices. The device controllers would only communicate with the cell controller, not with one another. The messaging would be over a MAP (Manufacturing Application Protocol) network using the MMFS (Manufacturing Message Format Standard) language (General Motors, 1984).

Given this set of specifications, the task was to design the control code that would reside in each of the controllers. The individual vendors would then implement the control code for their respective hardware, while we(Savoir) would implement the cell controller code.

Performance was certainly an issue. The overall assembly area had been designed hoping that each cell in it could do its job in 40 seconds. Kitting cell simulations had shown that under optimum conditions a kit could be produced in 40 seconds. The task looked daunting.

9.2.2 System analysis and design rationale

The simplicity of the kitting cell's operation is striking. Each aspect is straightforward to describe and easy to understand. However, like most real world control problems, a closer inspection uncovers numerous constraints and boundary conditions. Some of the most serious are as follows.

1. Even though there are numerous cameras, only one can be fired at a time.
2. The lane cameras shoot from above and can be obstructed by the part robot as it moves from lane to lane. The robot blocks three lanes to its left and two to its right while picking up a part. Picture-taking is an expensive operation; wasted pictures must be avoided.
3. Some of the cameras shoot two nests with each exposure. Those cameras should wait to shoot until both nests are filled with unanalyzed parts.
4. After a part is picked up, the nest must be given time to stabilize before the next picture is taken. The amount of time depends on the part in the lane.
5. Since the small lanes operate as a unit, turning the lanes on must be coordinated with the nest requirements and the small feeder operations.
6. Turning on the short lanes shakes any parts already in the three nests, thus invalidating any pictures currently in effect. So, if a part in a small lane is bad, dumping it and refilling must be considered in light of this constraint.
7. The gantry robot is a scarce resource shared across the entire assembly area. It is impractical (and selfish) to try to keep the lanes as full as possible. Instead, an algorithm must be devised which asks for gantry service as late

as possible, but before a lane runs out of parts. It would be nice if the algorithm were adaptive to the behavior of the gantry.

8. Restarting the cell from an empty configuration is hopelessly expensive. The cell must be able to recover without needlessly purging the lanes and feeders.
9. Similarly, reconfiguring the cell for different kits could require some lanes and feeders to be purged.
10. The kit and part robots could collide at the far right kitting station.
11. The long lanes cannot be stopped independently; there is one drive per pair of lanes.

There were other unsettled design decisions at the early control system design meetings. For instance, the camera could report a part as good, bad, or missing. The part reported as missing could be indicative of a lane being jammed or inoperative. In case of a jam, the part robot might be able to reach out and unclog the lane. An alternative was to increase the lane's amplitude to shake the parts free, recognizing that this affects both lanes.

No decision was made about what to do with kits that failed the final vision inspection.

A rack that holds kits was being considered. The idea was to decouple the kitting operation from the kit conveyor system. The rack could hold kits of various dispositions – empty, complete, partially completed (however they got that way), or bad kits (perhaps for later manual inspection). Presumably, the kit robot would know how to deal with them.

Other problems encountered in decision-making at these meetings were: the temporary bin storage space would depend on the exact cell layout, which would not be known until installation; the gantry protocol was unknown at this time; and finally, kit inspection took about seven seconds – perhaps the kit robot could move another kit into position in that time.

Since each device could be controlled independently, extraordinary parallelism was possible. However, it was exactly this flexibility that made finding an efficient control solution problematic. Parallel programming is difficult. The unsettled design issues only added to the design team's shaky foundation.

The design team started by generating a sequential process description which they hoped would lead to an implementation plan. The basic cycle was: initialize the devices; read the configuration information; make the gantry requests; load the lanes; move part robot to home; take pictures; pick up parts for pass 1; pickup kit from conveyor and put it on kitting stand; put parts in kit; pickup parts for pass 2; put parts in kit; move kit to inspection stand; inspect kit; put good kit back on conveyor.

The first thing they noticed was that this approach was inconvenient when considering the parallelism. (For example, loading the kit onto the kitting stand was completely separate from getting parts into the nests.) The next flaw came in dealing with the exception conditions; the flowchart that resulted was horrendous. Finally, the uncertainty of the unanswered configuration questions caused everyone to wonder if these meetings should wait until they were settled. Success of any kind became an issue.

At this moment we proposed an object-oriented approach. The intent was to build a prototype of the cell control logic that took into account all of potential design configurations. We were going to show how the methodology could create a control solution that itself had flexibility to change without drastic revisions to the control program. The solution was based on a object-oriented formalism we called Actors, which was to be implemented using the FLEXIS ToolSet.

The FLEXIS ToolSet

The FLEXIS ToolSet is a set of integrated computer-assisted tools that currently runs on a variety of industry standard workstations. It offers the following features:

1. Grafcet for building applications. Grafcet is a flowchart language for expressing control algorithms. FLEXIS has a graphical editor for constructing Grafcet charts. Within the Grafcet steps, the C or IEC Ladder language is used for data manipulation, logical testing, sending messages, etc.
2. Fully integrated messaging. Everything necessary to create, send, receive, and trace messages is part of FLEXIS. Manufacturing Message Format Standard (MMFS) messages on MAP networks are currently supported. The Manufacturing Message Service (MMS) (ANSI, 1989) will soon be supported in a limited way as well.
3. Graphical presentation tools, with which a variety of visual presentations can be built. Graphical and textual objects can be created and saved. Those objects can then be manipulated via a package of C functions callable from within Grafcet steps. The animation in Figure 9.1 was built using these tools.

FLEXIS is completely interactive during execution: the Grafcet is inspectable, it can be changed and incrementally compiled; the variables can be watched and its values changed; they can be cross-referenced; multiple windows can be opened; and so on. When applications are finished, they can be compiled completely into C and downloaded into selected runtime target hardware. The product overview (Savoir, 1988) gives a complete description of FLEXIS.

Figure 9.2 shows a Grafcet chart that has all the valid constructs. The boxes are steps that contain control code. Each step can contain three pieces of code: activation code that is executed once upon entry, deactivation code that is executed once upon exit, and continuous code that is executed while the step is active. The horizontal lines, called transitions, are boolean conditions. Steps stay active until the transition following them becomes true. Then control passes to the following steps. This scan based control architecture works as follows:

1. Evaluate all active transitions. A transition is active when all of its preceding steps are active.

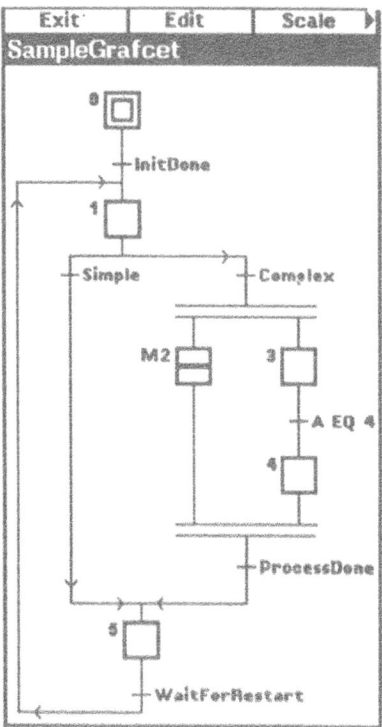

Figure 9.2 *A sample Grafcet.*

2. For those that are true, deactivate its preceding steps (running its deactivation code) and activate its following steps (running their activation code).
3. Run the continuous code in all active steps.

To start, the system activates the initial steps. From then on, the scan algorithm is used. This effect creates a time-sharing system across the grafcet steps. The designer breaks his control problem into pieces represented by steps and state-changing conditions represented by transitions.

Looking at Figure 9.2 in detail; it can be seen that a double boxed step, like step 0, is called an initial step. When the system starts executing, this is made active. Step 0 stays active until the variable InitDone becomes true. How that happens is immaterial. It may be associated with a sensor, a pushbutton, a message from some other process, or a timer. All that matters is its value. When it becomes true, step 1 will become active.

Step 1, which may be empty, waits for one of the two transitions following it to become true. This construct is called an 'asynchronous branch' because both branches could be taken if both transitions become true on the same scan. The construct following the Complex transition is called a 'synchronous branch'. It means that when Complex is active and becomes true, both steps M2 and 3 are activated.

Step M2 is an instance of a macro-step. It is just a collection of more Grafcet and offers the designer a convenient way to package up coherent chunks of Grafcet. Macro steps have a single IN and single OUT step. When the macro-step becomes active the IN step is activated. When the OUT step is active, the macro-step is ready for possible deactivation. Step 3 stays active until the variable A equals four.

When step 4 and macro-step M2's OUT step are both active, the transition ProcessDone is evaluated. This construct is called a 'synchronous join'. In a synchronous join all the preceding steps must be active before the transition is evaluated. It should be noted that step 5 has two paths into it from the transitions Simple and ProcessDone. This construct is called an 'asynchronous join'. Either or both paths can lead to activation of step 5.

Finally, step 5 stays active until the WaitForRestart condition becomes true and then branches back to step 1.

With these simple constructs any arbitrary control algorithm can be implemented. Because many steps can be active at once, Grafcet offers a language in which to express parallelism. But, note carefully, Grafcet by itself impose no methodology on the user.

Actors

Savoir has developed an object-oriented methodology, called Actors, for constructing Grafcet applications. Our Actor formalism offers solutions to controlling a collection of cooperating, parallel processes that act concurrently to do some job. The kitting cell is a perfect example.

In our methodology a problem is decomposed into areas of responsibility, each realized by an object called an 'actor'. The idea was to personify a manufacturing setting by placing independent actors on the floor to do the work needed to get the job done. These actors would communicate with each other naturally as they required. Like people, the independence of an actor forces a designer into conceptualizing the solution in terms of cooperating entities, not just one solution program. Thus, an application becomes a collection of these cooperating actors.

Each actor is a self-contained, object-oriented program. In particular, an actor is specifically designed to do work by carrying out one and only one task at a time. This feature makes actors simple to understand and control. Parallelism is achieved because all actors are running concurrently. The set of tasks in an actor obviously depends on its responsibilities.

An actor communicates with other actors by sending and receiving messages. The messages may request some service or supply status or any kind of information. Requests are typically stored in a queue.

An actor schedules its own tasks, including error recovery tasks, as required. Typically, scheduling decisions are made by processing a request queue, but often some decisions are made based purely on internal state. If no task is scheduled, the actor waits until some new state-changing event occurs.

An actor manages all the peculiarities of the physical device it may be driving. In our view physical devices are programmed to carry out specific actions which are driven by command messages from our actors.

Actors may be responsible for controlling physical devices, such as robots and lathes, for directing human operators, or for fulfilling purely logical roles, such as managing the quality assurance data of a factory. Some physical devices may have many actors assigned to them, if they have parallel capabilities. Importantly, each actor has the same architecture: a scheduler, tasks to run, actions that are carried out by physical devices, and messaging protocols for communicating with devices and other actors.

Objects in classical object-oriented systems are defined by class hierarchies with inheritance. The class defines a set of methods and associated invocation messages. Objects are created by instantiating a class. FLEXIS does not have such facilities. Instead, we have libraries of generic actors that are organized hierarchically. Users can copy actors out of these libraries and modify them as they see fit.

An actor is an autonomous player in a highly choreographed setting, performing tasks and interacting with a few partners through messages. This architecture is different from either a detailed orchestration centered around a conductor, or a structured top-down control solution. Wilczynski (1988; 1990) describes the actor design methodology further.

9.2.3 The kitting cell actor implementation

In this section, the actors in the kitting cell are described. The primary task and messages are given, actions if any, and a brief description. The message format shown below is informal. Parameters are shown in parentheses. In the actual implementation MMFS was used for all messaging.

The scheduling algorithms are not shown. Typically, they are simple: a message comes in and a task is invoked. Other actors, like the kit robot, must make priority scheduling decisions. However, they are unimportant to this discussion. The reader can easily intuit which messages are required in order to schedule the primary tasks. The distinction between tasks and actions attends to the real world of devices. Devices, like robots, lathes, and PLCs, all have special languages with special purpose interfaces to the mechanical things and sensors they are manipulating. The lack of standards is evident.

Abstractly, an actor is a logical entity, free of target machine details. Its tasks can be elaborated down to any level required. In practice, that level of detail, the mechanical interface, is fairly well understood. This means that an actor's task can be programmed down to the primitive level of sensors, elbow movements, and so on, if desired. Programmed this way, an actor could be downloaded into the target controller to drive the physical device. However, our actors are programmed in Grafcet, C, and IEC Ladder, and they send messages using a MAP model. Device controllers of the type seen on the floor rarely support any of these features.

So, in designing an actor controlling such a device, an arbitrary primitive layer is specified – the layer of actions. An actor's tasks interact with other actors through messages, and interact with its device by sending action invoking messages. The actors, so designed, can execute on a FLEXIS runtime, while the actions are all carried out in the devices. The actor task code bottoms out into actions, which are implemented in the device controller. Wilczynski (1990) elaborates on this technique.

Typically, the device vendor implements the actions. The actions and their associated invocation messages becomes a specification for him to implement. With this methodology, he only has to worry about carrying out his actions. He can ignore the problems of device-to-device communication and all the associated coordination issues. The actor is responsible for all that.

Kit delivery station actor

Primary task: Process kit – Get kit located in the station processed by the kit robot.

Primary messages: From kit conveyor – Kit placed (kit status)
To kit robot – Pick up kit (kit status)
From kit robot – Kit finished (kit status)
To kit conveyor – Kit finished (kit status)
Description: This actor manages the kit interaction between the conveyor and the kit robot. Notice how this actor insulates the kit robot from the kit delivery mechanism. This is a logical actor; no sensors on the kit delivery station, so there is no device control.

Kit robot actor

Primary task: Deliver kit –The robot picks up a kit from one station and puts it where it is needed.

Primary messages: From kit delivery station – Pick up kit (kit status)
From kitting stand(i) – Kit requested
To kitting stand(i) – Kit placed (kit status)
From kitting stand(i) – Pick up kit (kit status)
To kitting stand(i) – Kit removed
To kit delivery station – Kit finished (kit status)
Primary actions: Pick up kit
Move to location (x)
Place kit

Description: This actor is responsible for all kit movement. It knows how to handle a kit because of the kit status in the various messages: an empty kit goes on the kitting stand, a complete but uninspected kit goes on the inspection stand, etc. This status gives the cell flexibility in dealing with more than just empty and inspected kits.

Kitting Stand Actors

Primary task: Process kit – Get kit located in the stand processed by the part robot. At the end of the task a request is sent to the kit robot for another kit.

Primary messages: To kit robot – Kit requested
From kit robot – Kit placed (kit status)
To part robot – Assemble kit (kit status)
From part robot – Kit assembled (kit status)
To kit robot – Pick up kit (kit status)
From kit robot – Kit removed

Description: There are two of these actors, one for each of the kitting stands. The Initialize task will know if a kit is present, and so whether to prime the kit robot with a message or not. Notice how similar this actor is to the kit delivery station actor.

Kit inspection stand actor

Primary task: Process kit – Get kit located in the stand inspected by the vision system. At the end of the task a new request is sent to the kit robot for another kit.

Primary messages: To kit robot – Kit requested (complete kit)
From kit robot – Kit placed
To vision – Inspect kit
From vision – Kit inspected (kit status)
To kit robot – Pick up kit (kit status)
From kit robot – Kit removed

Description: This actor is virtually identical to the kitting stand actors except that it interacts with the vision system, not the part robot.

Long lane actor

Primary task: Deliver parts – This task manages the part activity for one of the lanes. It has two parallel responsibilities, one to interact with its feeder to keep parts in it, and one to manage its nest. Figure 9.3 shows the Grafcet for this task.

Primary messages: To feeder – Deliver parts now (number)
To feeder – Deliver parts if convenient (number)

From feeder – Delivery started
From feeder – Parts delivered (number)
From feeder – Parts not delivered
To vision – Part in nest
From vision – Nest bad (PartBadOrNestEmpty)
From part robot – Part removed from nest
To feeder – Dump nest
From feeder – Nest dumped
To feeder – Vibrate lane
From feeder – Lane vibrated

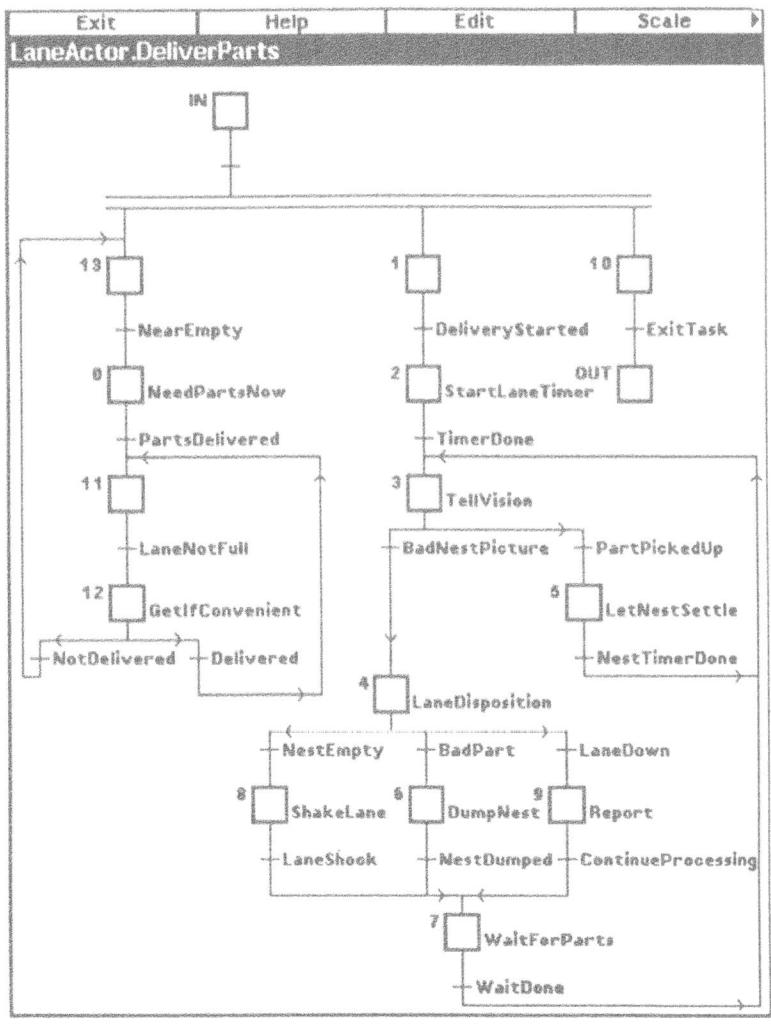

Figure 9.3 *Lane actor delivers parts task.*

Description: There are eight of these actors, one per long lane. Two of these actors will be interacting with one feeder actor. After the configuration task is complete, the 'Deliver parts' task will be scheduled. The basic 'Deliver parts now' message (step 0) will prime the lane. The 'Delivery started' message starts a timer (step 2) that tells it when a part should have arrived at the nest – there is no sensor in the nest.

Once parts arrive, a 'Part in nest' message is sent to the vision actor (step 3). If all goes well, the lane actor will next receive a 'Part removed' message from the part robot. The lane actor will immediately make a 'Deliver parts if convenient' request to its feeder (step 12). If the feeder has the right parts in it, the request is fulfilled. If the feeder has the wrong parts in it, it sends 'Parts not delivered'. The lane will wait (step 13) to make a 'Deliver parts now' request when it deems necessary; this is where the lane adapts to performance of the gantry robot and feeder in servicing its requests.

If the vision actor says 'Nest bad', the lane actor takes appropriate action: telling the feeder actor to dump the nest if the part was bad (step 6), or to vibrate the lane if the nest was empty (step 8). The feeder actor must coordinate these activities from its pair of lanes since either request could interfere with the other lane's activity. Step 4 can also decide that something is wrong with the lane and declare it down, after which a report is made (step 9).

This is one of the more complex tasks in the system. Its Grafcet exemplifies the value of a graphic finite state language. Trying to create such a graph for the entire kitting cell would, however, be horrific. Instead, we model each actor separately, decoupling their graphs, and use messages to coordinate them.

Short lanes actor

Primary task: Deliver parts – The task requests one part from each configured feeder. On completion, the lane drive is turned on for a predesignated time that will vibrate the parts to their respective nests. The task then interacts with the vision actor and part robot for the intended pickup.

Primary messages: Identical to long lane actor messages.

Description: Because there is one drive controlling all three short lanes, a single actor manages them all. The delivery task synchronizes the feeder delivery. Nest dumping and refilling is managed in a synchronized way as well. The messaging protocol here is identical to that found in the long lane actors, even though this actor is substantially different internally. The other actors have no idea that this lane actor is different from the others.

Feeder actor

Primary tasks: Load feeder – This task interacts with the bin manager to get parts of a certain type into the feeder. If the feeder needs dumping, the bin manager must be told to get an empty bin placed in the feeder's purge station.

Deliver parts – This task counts out parts for the lane that requested service. The task stays active as long as lane requests for parts match what it has in its feeder or until special services are needed. When a 'Deliver parts now' message comes that requires a changeover, this task will exist so the load feeder task can be executed.

Dump nests – This task dumps the nest. It is only scheduled when the lanes are in a appropriate state.

Vibrate lanes – This task vibrates the lanes for a pre-specified amount of time when the lanes are in an appropriate state.

Primary messages: From lanes – Deliver parts now (number)
From lanes – Deliver parts if convenient (number)
To lanes – Delivery started
To lanes – Parts delivered (number)
To lanes – Parts not delivered
From lanes – Dump nest
To lanes – Nest dumped
From lanes – Vibrate lane
To lanes – Lane vibrated
To bin mgr – Get purge bin
From bin mgr – Purge bin in place
To bin mgr – Load parts (type)
From bin mgr – Parts loaded
From bin mgr – No more parts

Actions: Start lanes
Stop lanes
Vibrate lanes
Dump nest (lane)
Configure lane for part (type)
Feed parts (number, lane)
Lower purge gate
Raise purge gate

Description: There are seven feeder actors. Four service two long lanes each, three service the short lanes individually. Despite this difference they are virtually identical. The long lane feeders must make decisions if it gets conflicting requests from its two lanes, e.g. 'Deliver parts now' from both lanes. That kind of complexity does not happen for the short lane feeders.

Bin management actor

Primary task: Get parts for feeder – This task had two parts: removing a non-empty bin from the feeder's purge station, and getting the right parts dumped into the feeder. The bin manager made requests to the gantry robot to carry out these actions.

Get purge bin for feeder – This task made sure an empty purge bin was put in the feeder purge station. Usually, one was there as a result of the 'Get parts for feeder task', but in case not, this task gets one there.

Primary messages: From feeder – Get purge bin
To feeder – Purge bin in place
From feeder – Load parts (type)
To feeder – Parts loaded
To feeder – No more parts
To gantry – Move bin (Loc1, Loc2)
From gantry – Bin moved
To gantry – Dump bin in feeder (Loc, Feeder, PurgeLoc)
From gantry – Bin dumped

Description: This actor maintains the database of bins located in the cell. It knows where the bins are, what is in the bins, and what are valid locations for the bins. How the bins were originally delivered was unspecified at this time; the bin manager isolated the cell from those unknown details. Only this actor would need modification when the bin delivery mechanism was determined.

The gantry robot would manipulate bins that were already in the cell. A protocol was invented since the real one was undesigned.

Gantry simulation actor

Primary task: Move bin – This task moves a bin from one location to another.
Dump bin in feeder – This task dumps a bin into a feeder and places the empty bin at the feeder's purge station.

Primary messages: From bin mgr – Move bin (Loc1, Loc2)
To bin mgr – Bin moved
From bin mgr – Dump bin in feeder (Loc, Feeder, PurgeLoc)
To bin mgr – Bin dumped

Description: This actor was purely a simulation of the gantry service that was eventually going to be put in place. We designed a simple, logical protocol until the real one was in place. The actor's task did little more than move the gantry figure around the animation. Simulation actors of this type frequently appear in prototype applications in order to 'run' them and see how they work. In the real application, the simulation actors are removed and if the messaging is the same, the application should just work.

Vision actor

Primary task: Fire lane camera – This task shot a lane camera and returned the results. Note that this task might satisfy two 'Part in nest' messages if the camera in question shoots two lanes.
Fire inspection camera – This task inspects the kit in the inspection stand and returns the results.

Primary messages: From lane – Part in nest
To lane – Nest bad (PartBadOrNestEmpty)
To part robot – Part good (lane, coord)
From part robot – You own lanes (lanes)
To part robot – You own lanes (lanes)
From kit inspection stand – Inspect kit
To kit inspection stand – Kit inspected (kit status)

Actions: LaneConfiguration (config)
Fire camera (number)

Description: The interactions among the vision, part robot, and lane actors are quite sophisticated. To review: even though there are many cameras, they can be fired only one at a time, picture taking is relatively expensive, some cameras shoot two lanes at a time, the robot moves across the camera's range of vision. All this means that a picture can only be taken when (a) the lane says a part is in the nest, and (b) the robot is out of the way. For cameras that shoot two lanes, it would be efficient to wait until the above conditions were true for the pair of lanes.

The air space above the lanes is a shared resource that the robot and vision actors manage together. Each is cognizant of what it means to 'own' the lane – the robot cannot cross the path of a lane owned by vision, and vision cannot shoot a lane owned by the robot. Vision gives a lane back implicitly with its 'Part good' message, and will give a lane back explicitly if the lane picture was bad.

Part robot actor

Primary task: Build kit – Each kit configuration and lane layout has one 'Build kit' task dedicated to it. The task implements a pre-designed pickup sequence across the lanes that tries to optimize the lane ownership interactions with the vision actor. It moves across the lanes picking up parts, telling the lane when it has, and giving up lanes to the vision actor when it seems appropriate. The kitting stand is informed when the kit is complete.

Primary messages: From kitting stand – Assemble kit (kit status)
                   To kitting stand – Kit assembled (kit status)
                   From vision – Part good (lane, coord)
                   To vision – You own lanes (lanes)
                   From vision – You own lanes (lanes)
                   To lane – Part removed from nest

Primary actions: Configuration (grippers, parts, lane data)
                 Pick up part (lane)
                 Put parts into kit
                 Move to rail position (pos)

Description: This is by far the most complicated actor. Its need to pick up parts is always tempered by its clumsiness in blocking so much of the vision's airspace. It has to subtly plan its trip across the lanes with the vision's work always in mind – both modelled carefully in the Grafcet task. When it wants to pick up a part in a certain lane it collects all the required lane tokens as they become available. As it moves across the lanes, it generally gives up lane tokens as they open up behind it. The teamwork here was elegant.

The performance of this actor and its robot was critical to achieving the overall cycle times. The typical Grafcet task in these actors has a step taking some action, generally by sending a message, and waiting on a transition for a reply before moving on to the next action. Here, that message is sent over a MAP network. At the time of this project, a round trip message took about 0.25 seconds. So, to send a pickup message, wait for the done message, send another pickup would be too slow; the robot would have noticeable delays between actions.

The problem was solved using a queue-ahead scheme. The robot was designed to be able to queue-up an action while executing the current one. Our actor knew about this capability and, anticipating success with one action, could queue up the next. This made the robot's movements smooth, without delays. Of course, this complicated the actor's code, but proved to be a critical technique also incorporated into both the kit robot and the vision actors.

Device simulation actors

Actors with actions send messages to the physical devices to carry them out. The feeders, vision, kit and part robots each were programmed in their own controllers. The actions, messages, and queueing behavior become the specification for the vendor. Our prototype also had an implementation of that specification purely to simulate the interactions to help check out the actors. In addition, the simulation drove the animation shown in Figure 9.1 so that the execution could be watched.

It worked simply. For example, the kit robot actor sends a 'Pick up kit' command message to the robot, not knowing or caring that it was destined for the real robot or the simulation. When the simulation was operating, the animated robot moved to the source station and picked up the kit. When the pickup was done the simulation sent a message back and the actor behaved accordingly. The animation was not a 3-D kinematic emulation, but enough to show what was going on in the cell.

Our object-oriented message-based architecture made it easy for us to stub in simulations of missing components.

Kit conveyor simulation actor

The kit conveyor system delivered kits to the cell. It was absent during prototyping and early integration so, like the device simulation actors described above, a simulation of it was built that implemented the messaging protocol. It tells the kit delivery station actor about the arrival of kits and removes them from the cell when told of their completion.

Factory control simulation actor

The factory control system sends configuration and control messages to cells, while receiving a variety of status and error messages in response. Like all the systems for which simulations were built, the factory control was offline during the early development of the kitting cell. The simulation tried to implement the factory control specification as closely as possible.

9.2.4 Bringing the cell online

The system as described so far has three key features:

1. The collection of non-simulation actors is the cell controller software. Because of the messaging interactions they can reside in any computer or computers that can deliver the messages.
2. The device simulations, which generated the vendor specification, will be replaced as the devices come online during the integration phase.
3. When the factory control, gantry, and kit conveyor systems come online, the kitting cell will be integrated into them.

This environment is not well suited for a 'design it, build it' mentality. The many unknown details suggest a prototyping one instead, where design and implementation happen continuously. To be able to see the 'whole' system operating gave everyone confidence that the project could succeed.

The prototype gave the mechanical engineers something concrete to evaluate. The roundtable design meetings all suffered from the vague verbal understandings that were being made. Watching the prototype execute in the FLEXIS system – the active Grafcet steps highlighting, the messages tracing, the animation moving – showed the engineers how the system would really work. Delays, complexities, and so on, were observed in the real control system code, not in hypothetical mathematical simulation models. When something worked well in our prototype, it figured to work well if the vendor delivered according to the well-tested action/message specification.

After some experience with the prototype was gathered, the vendors were given their specification. This separation of actor control code from the device action code was done carefully. The intent was to keep the device code stable; the programming environments in those controllers (especially the PLCs) were not conducive to checkout or reprogramming. We hoped any changes to the cell would happen in the cell controller part where FLEXIS offered a much friendlier and more powerful programming environment.

Integration

Of all the phases involved in getting a real-time manufacturing system into production, integration is the one most amenable to object-oriented techniques. Each actor has encoded into its tasks (a) all the coordination knowledge to interact with other actors; and (b) the actions and protocols for driving their device. The latter is the foundation for doing successful integration.

Special 'checkout-tasks' are added to each actor. These tasks are only used during the integration or maintenance parts of a cell's lifecycle. Their responsibility is to exercise the device and test the messaging interactions. These tasks comprise combinations of the actions programmed in the device. Even though the checkout tasks are completely different from production tasks, they use the same action code and messages, thus providing much-needed flying time for them.

For example, the kit robot actor had a checkout task that moves a kit back and forth between the inspection and kitting stands repeatedly. Though the task tested the calibration needs of the robot, it also exercised the primary pick-up and put actions, the action decoder, the robot device messaging, and the network. After this kind of testing, there was confidence that the production code was going to work.

The simplicity of this method figured heavily during acceptance testing of the vision system as well. After installation, parts were put in the lanes and pictures fired manually from the panel of the vision controller. The left-to-right sequence of pictures was replicated in the vision actor to see if the communications worked. When it did, all seemed well. However, using FLEXIS, a new Grafcet checkout task was quickly written that took the pictures in reverse order. Sure enough, the vision system failed and the acceptance was delayed until the problems were fixed. The nature of the FLEXIS tools, the autonomy of the actor, and the task/action decomposition all contributed to the discovery; no need to discuss the difference between a vendor's responsiveness to customer's needs before acceptance or after acceptance.

By having a rich set of actions, an actor can have a rich set of checkout tasks to test its device's fundamental activities. It is critical to check as much out as possible in a standalone mode because devices are not always delivered at once, or in the same state of readiness. Anything less is indicative of poor methods. By the time we have done this testing to our devices, there is good reason to believe they will work in production.

As more devices are brought on-line, more interactions can be tested. The robot and vision actors' interactions were tested this way. By running simplified versions of the kit building tasks, many metrics were gathered that helped improve their cooperation.

Performance

The system performed elegantly. With approximately 100 conversations going on simultaneously it seems miraculous that anything can get accomplished. However, each individual conversation is simple and logical. Each actor behaves rationally because of the intelligence built into its scheduler. Each task is meaningful given the circumstances of its invocation. And despite the messaging overhead, the overall performance was acceptable – a kit could be produced in about one minute. Any loss due to the multi-processing and messaging overhead was gained back by the effective parallel utilization of the devices. In fact, despite being the most complex of the cells, the kitting cell had the fastest cycle time of any in the assembly area.

Compared with standard top-down or structured solutions, this system of actors was superior in every way. The object-oriented methodology makes a system's complexity linear instead of exponential with respect to the number of devices under control. It is this last feature that was so critical during the ongoing evolution and maintenance of the cell.

The overall system was reliable and understandable because each actor was reliable and understandable. Tuning the system meant tuning individual actors, not a single large body of code – a feature that gave confidence to the engineering staff when considering improvements to the cell.

Maintenance/improvement

Throughout the project, the mechanical engineers were encouraged to modify the working of the cell without regard to the cell control program. We were sure that any changes they made could easily be incorporated into our actor architecture.

The first change they made was to replace the kitting stand table with an automatic index table. The new table had three positions, a load position, a kitting position, and an inspection position. Indexing meant moving the kit from the load to the kitting position, from the kitting to the inspection position. The indexing rules were obvious. The kit robot put empty kits into the load position and removed kits from the inspection position. This addition simplified the kit robot's job and slightly improved the overall cycle time. At startup time, an empty kit was put in the load or kitting position.

For the actors, it meant replacing the kitting stand actors with a new index table actor. This new actor behaved identically with respect to the two robot actors, so they needed no change. Its main processing task did the work of coordinating the kit activities. It had one action in the PLC to do the indexing.

An improvement was made to the kit robot actor. The preconditions of its 'Deliver kit' task demanded a kit to be available and a stand on which to put it – it would not pick up a kit unless it had a place to put it. Now with the indexing table there was little chance of getting stuck holding a kit while something more important had to be done. So the precondition was weakened. When a kit appeared at the kit delivery station the task was started in anticipation of the indexing, which freed up the load position. A simple improvement and trivial to make.

The part robot actor needed many improvements. It jammed quite frequently putting a certain part into the kit. It was difficult to fix the mechanical problem. Instead, automatic error recovery was added to the actor for this particular problem. The context of the kitting task together with the error message made the problem easy to decode.

To speed up the performance of the part robot, two-level queue-ahead was added. This required a small change in both the robot controller and the actor. This extra queue level meant the robot could be holding two actions while executing a current one. By watching the robot, the messages, and the Grafcet kitting task, we could see that by the time the actor received its acknowledgement of a pickup, the robot had already finished both of the queued pickups as well. The actor was modified to anticipate the robot's movement in order to free the lanes more quickly for the vision system. It is precisely this kind of tuning that is possible in object-oriented systems.

9.3 The glass line

9.3.1 Specification

In the kitting cell an object-oriented approach led to a successful cell control implementation. The glass line was already operational when a similar analysis was sought. An offline model was built with three goals:

1. to enhance throughput by improving the conveyance algorithms;
2. to eventually supply a full control program for the line; and
3. to show that part tracking is feasible and could reduce many of the manual responsibilities.

The glass line, shown schematically in Figure 9.4, turns raw glass panels into windows and windshields for automotive vehicles. Glass-finishing processes are

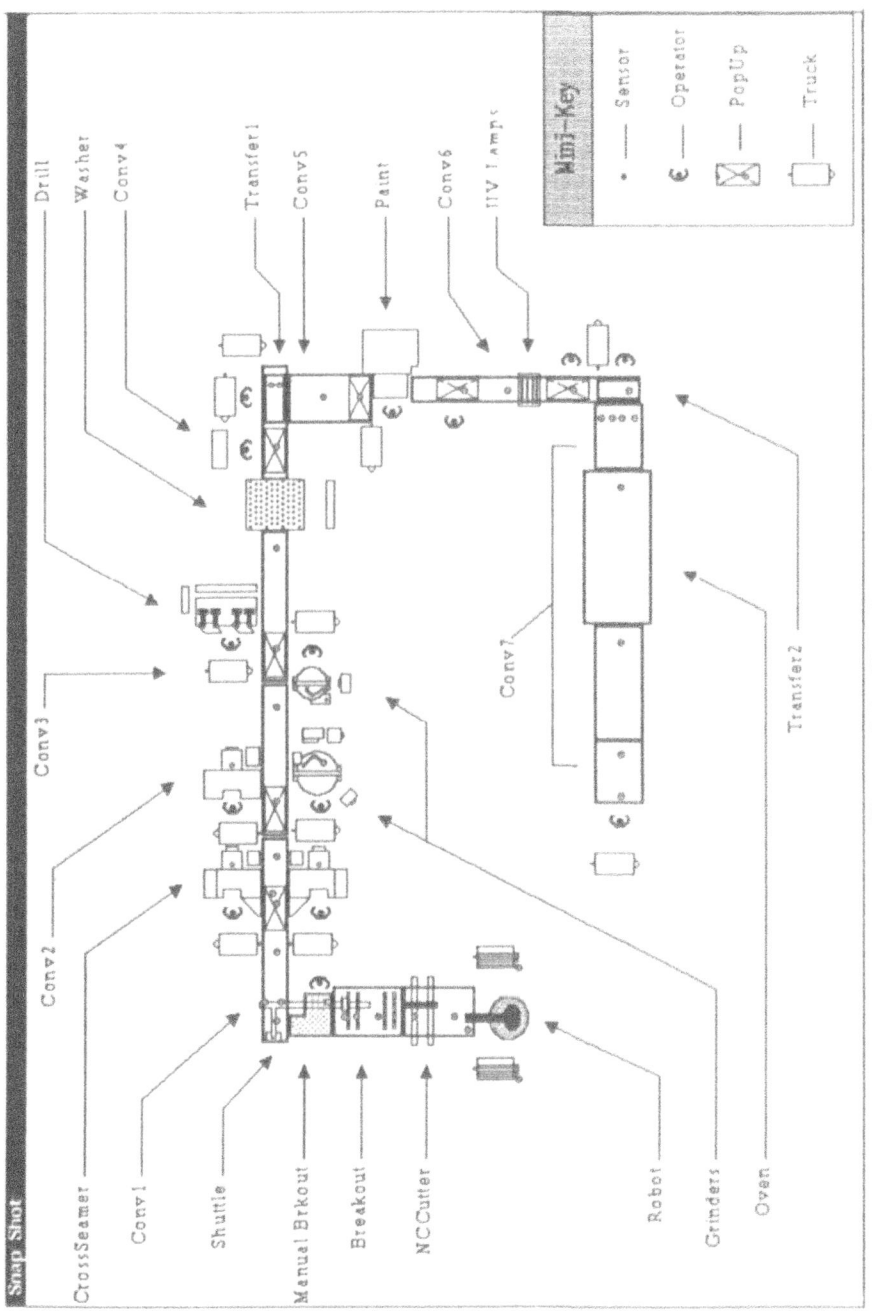

Figure 9.4 *The glass processing line.*

linked together by a segmented conveyor system. Some of the processing is done automatically right on the conveyor, but most happens as operators move the glass between the conveyor and their workcenters.

The line is in a laboratory developing glass manufacturing technology. New glass processes are continually being tried and evaluated. The line is continually being reconfigured to accommodate them. However, the line is also a working prototype with daily production quotas and an intention to be a model for all the company's glass lines. The desire to continually improve throughput is strong. Though the glass-processing is the primary focus, the line management is almost as important. It is here that the object-oriented analysis was expected to yield its main results.

The glass line's basic cycle has four parts:

1. A robot picks up a glass panel from either of its input bins and puts it on the first conveyor (NC Cut Conveyor). From there the glass moves into position where an NC machine cuts a template into it. The specific NC part-program is dialed in manually. The operations are keyed by sensors on the conveyor. After cutting, the glass is heat-treated, and the template is broken out of the glass. The shuttle automatically moves the template onto conveyor 1. If the breakout was unsuccessful, an operator can manually finish the process.

2. At the top, four conveyors move the template through various phases of edge-treatment. Not all operations are necessary on all templates. Each of those conveyors has a pop-up table (the boxed X) which will raise the glass on it (unless the operators signal otherwise by pushing a bypass button). The operator then slides the glass off the pop-up into his work-area for specialized processing. When the operator is done, he slides the glass back onto the raised pop-up and pushes a button to lower it. He can raise the pop-up directly as well. The conveyor can be moving while the pop-up is raised.

3. On the right side, conveyors 5, 6, and 7 move the template where it is manually painted and then treated with ultra-violet light. The same pop-up mechanism is used here when manual processing is required.

4. Finally, on the bottom the paint is baked on in a special oven. The glass template is manually removed at the end.

The line is controlled by three PLCs programmed in their native relay logic ladder language. All the conveyor and pop-up logic resides in one of them. The object-oriented analysis will be most important here.

Each conveyor has combinations of the following controls and sensors: start/stop conveyor, glass detection sensors, raise/lower pop-up (which can be enacted by operator push-button), pop-up sensor, pop-up bypass switch (enacted by an operator push-button). The placement of sensors and pop-ups varies from conveyor to conveyor. Typically, there is a sensor preceding a pop-up and one on the pop-up. Each conveyor has an exit sensor, which sometimes doubles as the pop-up sensor if the pop-up is at the end of the conveyor.

9.3.2 System analysis and design rationale

The main goal of the system, to keep glass flowing efficiently around the line, is tempered by the danger of glass panels crashing into one another or into pop-ups. This kind of continuous flow system is difficult to manage using standard methods. Like the kitting cell with its many devices and hundreds of potential interactions, the glass line has many conveyors to coordinate with information coming from dozens of sensors. The continual reconfiguration caused by the glass-processing experiments only added insecurity.

The technical response was to make conservative control decisions. For example, on conveyor 1 only one part was allowed to approach its pop-up, even though there was room for several; a new part could be loaded onto the conveyor only as a part was leaving the pop-up.

In the original implementation of this decision, a part could approach only if the pop-up was down. Otherwise, the conveyor would stop and wait for the pop-up to lower. When it had, the conveyor would move the part off the pop-up as the next part approached. When the trailing edge of the processed part triggered the pop-up sensor, a timer would start. Its expiration signaled that the new part had arrived and the pop-up could be raised.

This solution required a pre-designed flow of parts on the conveyor. The approach times were hard coded into the algorithm and the operators had to know the loading rules. Though simple, it was inefficient. By ignoring most of the available sensor information, the implementation was unreliable as well. Operator and timing errors were undetectable. Unfortunately, coding simplicity won out over robustness.

The conceptual difficulty of managing so many connecting conveyors, each with slightly different layouts and slightly different entry/exit requirements was daunting. The inherent volatility of the line's overall configuration just added to the problem. Yet it is precisely this type of problem that yields to object-oriented methods.

In our solution model, a conveyor segment is divided into zones each having at least one sensor. Each zone will be managed by an actor. The zone actors are responsible for getting parts into them, doing any special glass processing if required, and getting the parts out of them. Each conveyor segment also has an actor to resolve the zone actors' potentially conflicting start and stop requests. Thus, each particular conveyor segment is managed by a 'community' of these actors. These communities, in turn, interact with one another because one's entry actor interacts with the previous one's exit actor. The result is a large set of cooperating actors managing the complete line.

Looking at the glass line schematic in Figure 9.4, conveyor 1 has an entry zone, a pre-popup staging zone, a pop-up zone, and an exit zone. Each will be assigned an actor. The entry actor will interact with the shuttle, while the exit actor will interact with conveyor 2. Conveyor 1 will also have a start management actor, hence a community of five actors.

9.3.3 The glass line actors

The actors fall into three categories: load, stage, and start. In theory, each would be a class. A community for a particular conveyor would be just a collection of the required actor instantiations. Each instantiation would be modified to identify its neighboring actors. Specialized glass processing would be added if required.

In FLEXIS Grafcet, there are no classes. Instead we build macro-step libraries. Instantiation means copying the library macro-step and making the required modifications. Here are descriptions of the basic actors.

Glass line load actor

Load task: If empty, it sends a ready message to the actor who precedes it on the line. When a part is detected in its zone, it tells the start manager to turn on the conveyor and waits for the part to leave.

Primary messages: To preceding actor – Ready for parts
                           To start actor – Start conveyor

Action: Stop conveyor

An entry actor is just an instantiation of the basic load actor. There are, however, two specializations of load actor to consider. They process the part in its zone before moving it on:

1. Pop-up actor – This load actor does pop-up management once a part arrives in its zone. It will have additional actions for raising and lowering the pop-up.
2. Processing actor – This load actor signals for the on-line automatic glass processing when the part is detected in its zone.

Glass line staging actor

Staging task: When a part is detected, the conveyor is stopped until the next actor in the line signals ready. When he does, the start actor is notified to start the conveyor.

Primary messages: From following actor – Ready for part
                           To start actor – Start conveyor

Action: Stop conveyor

An exit actor is just a simple instantiation of a staging actor.

Glass line start actor

Start task: If the conveyor is stopped, check the start messages of its conveyor community. When they all say start, the conveyor may be turned on.

Primary messages: From zone actors: Start conveyor

Action: Start conveyor

9.3.4 The glass line actor implementation

The full glass line implementation is a large collection of actors and simulations. There is a community of actors for each conveyor segment: NC cutter conveyor, breakout conveyor, conveyor 1, conveyor 2, conveyor 3, conveyor 4, conveyor 5, conveyor 6, and conveyor 7. There is an actor for each on-line device: NC cutter, breakout, washer, ultra-violet lamps, and oven. There is an actor for the robot and shuttle. And there are simulations for each of these actors as well as for all the manual operations around the line.

Because of customer requirements, shared variables were used instead of explicit messaging. However, the communication variables were carefully identified and managed, as were the variables belonging to each actor. The logical encapsulation was maintained.

Except for the conveyor actors, the others are much like those found in the kitting cell. The major difference appears in the real-time needs of the conveyor communities and in accurate simulations for study. The conveyor 1 community is typical of a conveyor segment.

The conveyor 1 community

Figure 9.5 shows the Grafcet for the conveyor 1 community. Conveyor 1 has four zones each represented by a Grafcet macro-step. The fifth marco-step, StartMgmt, completes the community. Each of these macro-steps is a complete, though simple actor.

The Entry macro-step, an instantiation of a load actor, is shown in Figure 9.6. Once entered, two steps are made active: one listens for Reset in order to exit; the other begins the work. This structure is part of each actor. Step 1 waits until the entry sensor, linked to variable C1PartAtEntry, is off. With its zone empty, the actor asks the shuttle for a part (activation code in step 2) and then

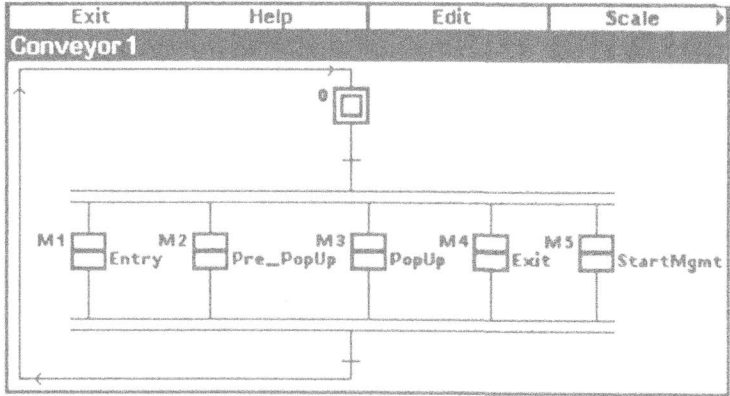

Figure 9.5 *The conveyor 1 community.*

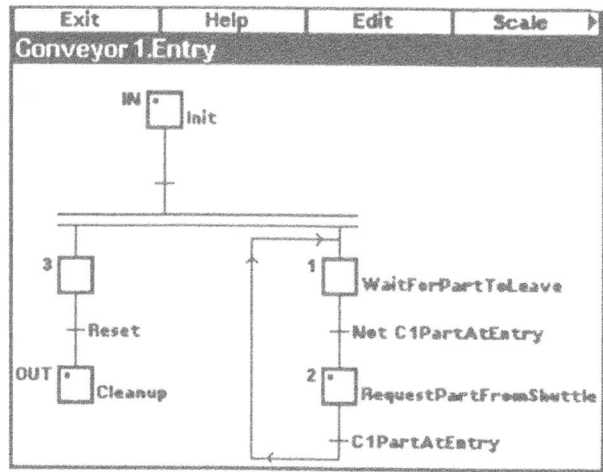

Figure 9.6   *Conveyor 1 entry actor.*

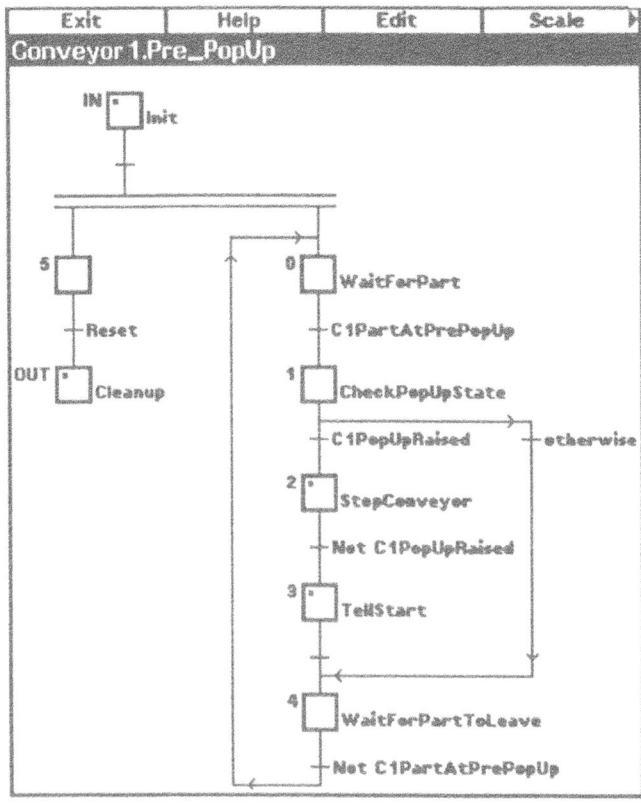

Figure 9.7   *Conveyor 1 pre-popup actor.*

waits for a part to appear. Note that this actor never stops the conveyor and, hence, never has to signal for it to start. All its logic is based on the part present sensor.

The Pre_PopUp macro-step, an instantiation of a staging actor, is shown in Figure 9.7. It starts by waiting for a part to arrive at its sensor (step 0). When it does, step 1 checks to see if the pop-up is raised. If not, it waits (step 4) until the part leaves the zone. If the pop-up is raised, it stops the conveyor (activation code in step 2), thus avoiding a collision. When the pop-up is lowered, the actor tell StartMgmt to start the conveyor (activation code in step 3). It then waits (step 4) until the part leaves the zone.

The PopUp macro-step is shown in Figure 9.8. It waits (step 0) for a part to appear at the pop-up. When one does, it raises the pop-up (macro-step M2) unless the bypass mode was set by the operators. Once the pop-up is raised (macro-step M2), it waits (either step 4 or macro-step M2) until it is lowered, after which it waits (step 3) for the part to leave. Notice that the conditions under step 0, C1PartAtPopUp and C1PopUpRaised, can both be true. In that case both steps 1

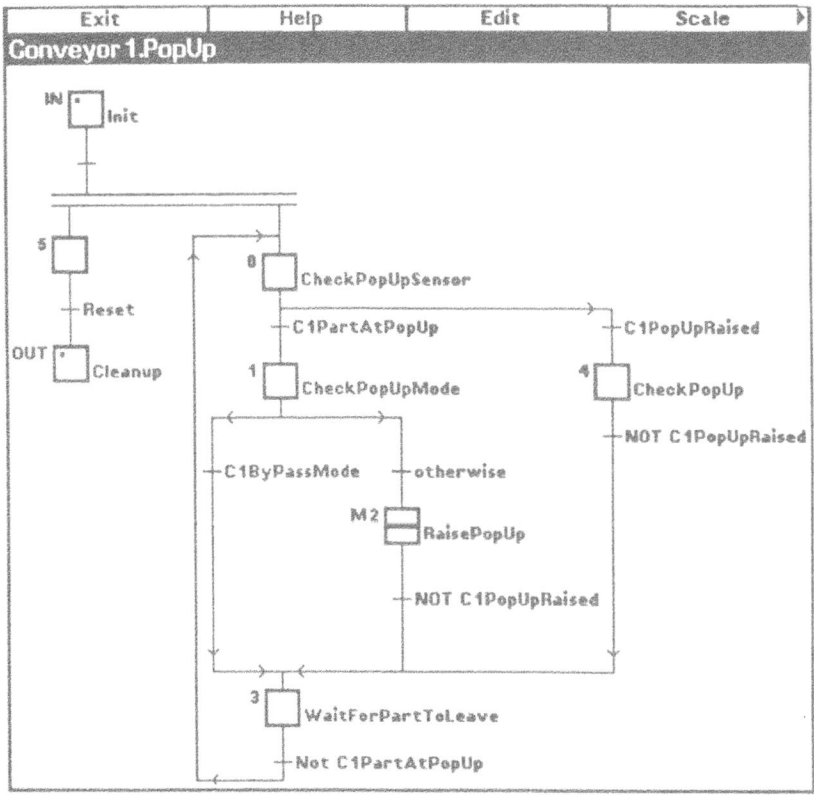

Figure 9.8 *Conveyor 1 popup actor.*

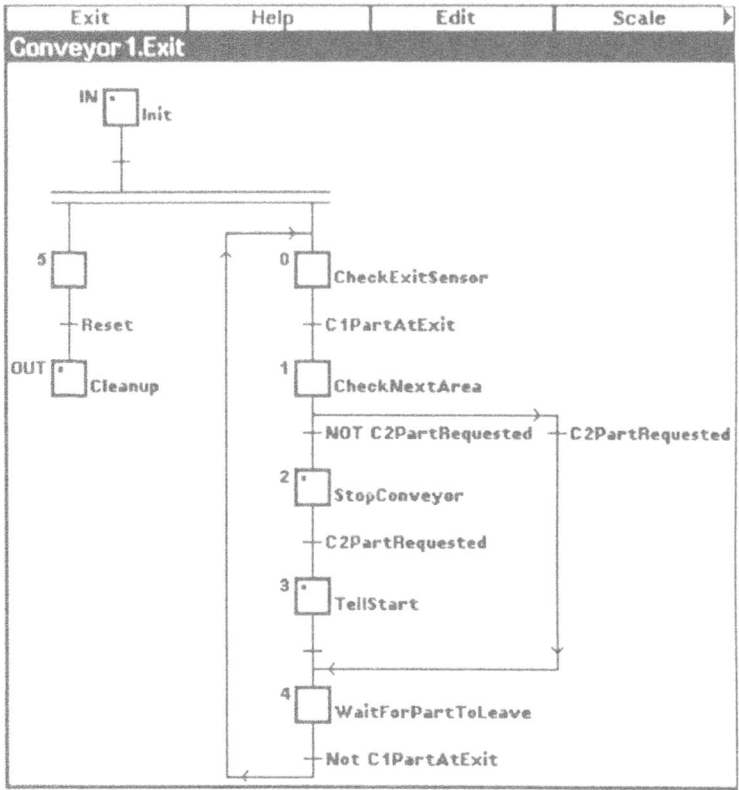

Figure 9.9   *Conveyor 1 exit actor.*

and 4 will become active. A straightforward analysis shows that these two paths will synchronize at step 3.

The Exit macro-step, a kind of staging actor, is shown in Figure 9.9. It starts by waiting for a part to reach its sensor (step 0). It then checks if the conveyor 2 is ready for a part (step 1). If so, it sends it on, waits for the part to leave (step 4) and then repeats its cycle. If conveyor 2 is not ready, the conveyor is stopped (step 2) until a part is requested. Then, it asks StartMgmt to start the conveyor. As before, it then waits for the part to leave (step 4) and it repeats its cycle.

The StartMgmt macro-step is shown in Figure 9.10. It waits for the conveyor to be stopped (step 1). Then, it starts to check the community for a start consensus (continuous code in step 2). When all agree, it starts the conveyor (step 3) and repeats its cycle once the conveyor is running.

The simulations

Rapid prototyping and offline testing are part of the FLEXIS philosophy. It is important to execute the control code before attaching it to real devices. In the

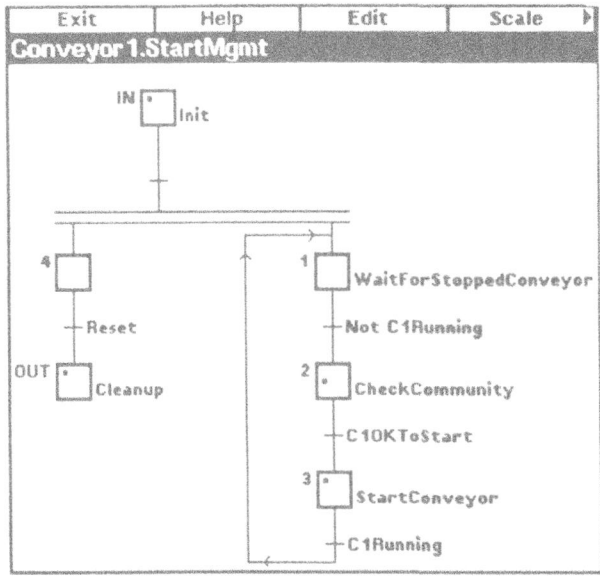

Figure 9.10 *Conveyor 1 start actor.*

kitting cell, the simulations were rather coarse; they were at the action level only. The animation was driven by the action level calls, while lane and feeder details were relatively unimportant. The opposite is true of the glass line. A throughput analysis was primary, thus conveyance simulation was key. The object-oriented techniques served us well here.

Each actor is associated with its own device simulator. That simulator manages all aspects of the 'real-world' being simulated. The control actors simply respond to events, whether they are generated by simulations or the real devices. The simulations fall into three categories: discrete-event, part movement, and sensor triggering. The overall simulation is depicted in an animation.

Actions are modelled to take a certain amount of time. For example, a robot MovePart action may take five seconds. After the action is started, the simulation reports the completed event when five logical seconds have elapsed. This discrete event management is done classically. Actions are put on a global event queue. As a simulated clock advances, expired events are reported. The simulator does this by counting Grafcet scans and computing an elapsed time based on how long the engineers feel a scan should take.

The part movement and sensor detection are more difficult to simulate. They are managed at the conveyor community level. A model of the conveyor length, conveyor speed, part size, and sensor placement all contribute to accurate simulation. When a conveyor community is laid out, an appropriate set of simulation tables must then be filled in. The part movement simulator can then generate realistic part movement and sensor activation.

The simulation was a critical part of this project. The engineering organization needed to be shown that control models could be built and tested offline. Typical time-compression discrete-event simulators are fine for gross analysis, but whether those models reflect the reality of their after-the-fact implementations is another thing. Our approach ties the simulation directly to the control models. The analysis is done directly on the Grafcet implementation, so there should be more confidence that the offline model will in fact model the reality. In fact, as will be shown later, the statistics generated by this model were fed back into a time-compression discrete-event simulator for further analysis.

9.3.5 Results and benefits

Configuration flexibility

The system as describe above now had the flexibility originally desired by the glass engineering staff. The conveyors were simple to configure given the actor decomposition, and new glass processes could be introduced with the confidence that the line would still work.

The whole notion of having this offline model to experiment with fulfilled the main customer objective. Manufacturing typically has had to install a system to see how it works. On-the-floor debugging is difficult, on-the-floor continuous improvement is impossible.

Performance, improvements, and experiments

The result is remarkably robust and efficient. As is typical of object-oriented systems, there is no hierarchy of control. All the actors are peers with different and limited responsibilities. The independence of the actors will allow almost arbitrary reconfiguration to the line, yet with confidence in the safety of the line.

The accurate simulations allow the engineers to play 'what-if' games: moving sensors around, adding conveyors stations, and so forth. The utility of such changes can first be tested with the model.

Several such experiments were actually implemented. The communities for conveyor 1 and conveyor 5 were recoded into the conveyor PLC. Thus multiple parts could now approach the pop-ups. The original timer-oriented algorithms with their operator restrictions were replaced by the sensor-driven actor implementation. Sensors were also repositioned based on experiments with the model.

Other experiments were tried that were not implemented. One involved adding a pop-up at the entry of conveyor 1. Presumably, it would help the operator in manually loading glass templets. The pop-up allows the operator to transfer the part without having to stop the conveyor.

Another experiment tried configuring conveyor 6 into smaller segments. The advantages of adding another paint conveyor for improved throughput was tried as well. The point is that these experiments were possible to do before on-the-floor engineering decisions had to be made. The object-oriented methodology made it possible.

Part recipes

Interestingly, most of the line decisions depend on human recognition of what was going on. Nothing in the line's automation knew what kind of parts were flowing through. An operator had to dial in the NC part program number for the glass parts being loaded by the robot. Each operator at a glass-processing station had to declare, using the bypass buttons, whether or not he wanted the part approaching the pop-up. The parts were moving around blindly, totally dependent on the operators to manage the interface between stations. To simplify things, templates were done in large batches; different types of templates were not mixed together on the line.

In our object-oriented model, where parts are essentially handed off between actors, part type information could be handed off as well. The actors could make decisions on their own, making it easier to mix parts on the conveyor.

The model was changed so that at every entry or exit station, an operator can tell the executing model about parts entering or leaving the system. For entering parts, he must tell what kind of part it is. Though the information was typed into the experimental model, bar code readers would typically be used for production. Most of this part data entry happens at the robot station.

The system has part recipes that declare what must be done to each type of part. As parts move from conveyor zone to zone, the part type information is passed between the actors along with updates about the state of the part. This scheme makes more automation possible. For example, the NCCutter part programs can be downloaded directly by the actor instead of it being dialed in by an operator. Similarly, pop-up actors can tell if the part is to be processed at its station; the operator need not push the bypass buttons to give the system that information.

Timestamp information was stored as well. These statistics were important for the process engineers to study, but they were also used by other analysis packages. The statistics were fed into a time-compression discrete event simulator (not to be confused with the actor simulations discussed above). Suddenly, it was possible to study the different patterns relating to the part mix, changeovers, start-ups and shut-down conditions, etc. The new data was fed into a time-compression discrete-event simulator in order to generate throughput statistics over long periods of time.

9.4 Conclusion

The kitting cell and the glass line benefitted from a prototyping mentality made possible by object-oriented techniques. Both had processing models that were

226 OOPS in real-time control applications

fairly easy to state and understand, yet were difficult to implement as control solutions using classical techniques.

The basic problem is two-fold: (a) systems are hard to understand until they are implemented and (b) once systems are implemented, they are hard to fix or improve. The former implies a need for rapid prototypes, the latter a need for an implementation that is flexible, understandable, and changeable. These case studies showed how both problems were addressed using object-oriented techniques.

The kitting cell showed how the prototype led to understandings about the device interactions and suggested improvements in the cell's mechanisms. The prototype evolved with those changes; without the encapsulation offered by the actor methodology, the software could never have kept up with the improvements. The glass line model did exactly the same thing, but here the line was already implemented and operating poorly.

Prototyping and understanding are brothers under the skin. The kind of models we built here created a foundation for understanding and improvement. This kind of analysis has typically been done using only time-compression discrete-event simulations. However, those models are never elaborate enough. Things turn up during implementation that were missed during the simulation. The glass line work showed how FLEXIS and a discrete event simulator can be used together.

The FLEXIS implementations are concrete models intended to control the target system. The systems can be prototyped quickly, analyzed, and improved. Then, if the implementation is used for control, integration is straightforward, as it was in the kitting cell. Or, if the model is used to improve a process, some of the algorithms in the prototypes can be reimplemented in the actual system, as they were in the glass line. Object-oriented methodologies have a natural place in manufacturing; successful case studies will demystify their use.

References

ANSI (1989) *Manufacturing Message Specification (MMS)*, EIA Standard 511, American National Standards Institute.

General Motors (1984) *MAP Specification 2.1, Appendix 6 (MMFS)*.

Savoir (1988) *FLEXIS product overview*, Hayward, CA.

Wilczynski, D. (1988) A common device control architecture – The Savoir actor. Proceedings of *Autofact '88*. November, Chicago, IL.

Wilczynski, D. (1990) Creating a control system application – From concept through production. *Proceedings of IPC '90*, April, Society of Manufacturing Engineers, Drarborn, MI.

Summary: Part Three

S. ADIGA

Two chapters in Part Three show three significant applications of object-oriented software. Barry Lozier of Consilium presents an ambitious plant-floor management system; David Wilczynski and David Wallace of Savoir present a detailed view of two applications in real-time control areas.

Flowstream: A new plant-floor management system by Consilium

The main advantages sought by Lozier's project were reusability, portability and modularity. Keeping these objectives in sight during the progress of the project seemed to help Consilium make key trade-off decisions involving OO software.

In a large project, cross-references by a large number of different objects from different classes may lead to a 'spaghetti-like' structure. Consilium organized object classes into major groups that are partitioned into 'layers'.

Another key element appears to be the interfacing of their system with a relational database. This allowed them to take advantage of the proven strengths of the relational data model, and focus more on gaining OO benefits in application software development. This may also be seen as a way of minimizing the risk taken by the company in adopting a new technology early (by mixing it with known technology).

The literature is full of papers complaining about the learning costs involved in changing over to OO development. It is interesting to note that Lozier claimed that the cost of learning a new paradigm has been more than offset by the ability to perform new upgrades and move to new hardware platforms. His paper contains many lessons for software developers intending to switch to the OO platform for product development.

Real-time control applications by Savoir

Wilczynski and Wallace discuss two very impressive applications in great detail. The first one is a new project; that is, they use OO techniques to build a prototype using their reusable software tool kit, FLEXIS, to study the mechanical interactions and system behavior of a new system being built. The second application addresses issues involved in improving an existing system.

They present an interesting design approach that starts by decomposing the problem into areas of responsibility. The latter are then assigned to 'active' objects called Actors. Each actor has its own thread of control, therefore it can act independently of the others. The coordination is done through message-passing protocols.

One of the problems facing implementors of real-time control software is that of interfacing with different protocols of various vendors of equipment or control devices. Wilczynski and Wallace handle this problem by separating the device-independent communication from the device-specific communication. This ensures portability and demarcation of responsibility for the hardware vendor and software vendor. The authors achieve this separation making a distinction between tasks and actions within each actor. Tasks trigger actions which in turn send appropriate messages to the devices to be controlled.

Flexibility in this approach, by encapsulating real-life devices within objects and communication through messaging, allowed Savoir to cope with a project that had unsettled design decisions.

Wilczynski and Wallace describe their problem and design rationale in much detail. We feel that the example contains many ideas on modeling a cell control system effectively. Perhaps their the most important contribution is their convincing demonstration that it is possible to tune the design of a target control system by observing the performance of an OO program. Tuning a system includes decisions such as: where do you need error recovery, where to inject human intervention, which operations to be optimized, where to reduce message traffic? etc.

The authors' experience suggests that a real-time control application can conform to the OO paradigm by building a simple prototype that models the basic functionality of the target system and conservative control logic. More detailed and complicated requirements such as error recovery, making use of sensory information, and speeding-up of performance can be added gradually. The natural correspondence between simulation objects and real-life devices helped relate the changes quickly.

The first example dealt with the improvement of performance of a prototype. The second example takes it further by tying simulations directly to control models in a working system. We would also like to remind readers that prototyping is not only an exercise in building software for the domain but also an exercise in improving one's understanding of the real-life system.

Concluding remarks

Lozier has presented a commercial software vendor's development perspective, while Wilczynski and Wallace have emphasized issues related to implementation at the user's site. The latter show that OO methods fulfil the needs of real-time manufacturing naturally and efficiently in their two examples. Together, the two chapters address a wide range of issues concerning the commercial development and deployment of OO software. These chapters are aptly complemented by Adeel Najmi in Part Four, who presents the organizational and management aspects of introducing and nurturing OO software within a manufacturing division.

Part Four
Management and
Organizational Issues

10 Management issues in adopting object-oriented technology

ADEEL NAJMI

NCR Microelectronic Products Division, Fort Collins, Colorado

10.1 Introduction

We first became involved with object-oriented software technology for manufacturing in 1988 within a project to develop a manufacturing scheduling tool (Najmi and Lozinski, 1989). We were attempting to automate a production activity forecast that had been done manually by a human scheduler for a few years. We realized early in our knowledge engineering effort that our expert was essentially doing a mental simulation of the manufacturing activity in order to compute the forecast. We needed a way to represent the expert's knowledge of factory data (such as throughput times of operations) and behavior (such as task execution behavior and decision-making heuristics) in a computational model that could be used to simulate detailed production activity for our semiconductor wafer fabrication facilities. We found a convenient starting point in an object library (Glassey and Adiga, 1989) developed at a university to support production scheduling research. Thus, our first application was developed by reuse and extension of an externally supplied object class library.

Use of an existing library allowed us to deploy a prototype in as early as three weeks of software development. This rapid prototyping ability allowed us to work closely with our expert and users throughout the development period as we incrementally refined our system. We found it helpful to be able to communicate our abstraction in software using the object-oriented paradigm. We were able to release our system to full production use in five months with a development effort of only eight man-months. Later, this system was also installed at a second wafer fabrication plant in 1989.

Over the course of the next few years we have continuously extended this system by addition of functionality, refinement of the model, and improvement in the decision-making heuristics. In 1989, a second application was developed in response to similar needs for our wafer probe facility (Adiga and Lin, 1990). In this second application we added the ability to incorporate rule-based inference in order to implement certain heuristic knowledge related to setup change decisions. This application was built entirely by subclassing objects developed at a university. In another project, in the summer of 1990, we added the capability to explicitly represent labor resources in our models. In 1991 we extended our wafer fabrication model to represent kanbans and their behavior within a 'just-in-time' manufacturing program. In future, we expect the number and complexity of our object-oriented applications to grow further in response to user needs.

Interest in object-oriented technologies has now become quite widespread in the industrial arena. Many companies have already embraced these technologies within their software development and research groups. Adoption of object-oriented software development technology involves much more than a mere change in software development language. The issues are both managerial and technical. These issues require a conscious paradigm-shift for the entire organization.

Due to the lack of practical design methodologies that are supported by CASE tools, we had to develop our own methodologies. Also, as with most other groups that adopted object-oriented technologies early, we had to learn our own management methods during the course of our projects. On hindsight, there are perhaps many common lessons that were learned from different projects. This chapter reflects our experience from the projects discussed above as well as some insight gained from observation of other development groups both inside and outside NCR.

The organization of this chapter is as follows. We first describe distinct differences in the software development lifecycle experienced in our object-oriented projects and the lifecycle of conventional software development projects. We then discuss issues of reusability, resourcing and supplier management. Given this context we discuss planning and budgeting for such projects. Finally we present some concluding remarks.

10.2 Object-oriented software development

10.2.1 Project lifecycle

While the model developers are busy attempting to propose elegant new models, software development needs to continue to proceed in industry and cannot wait for the dust to settle. Traditional models of software development project lifecycle consist of the following phases:

1. requirements analysis;
2. external design;
3. internal design;
4. implementation (coding);
5. testing and validation;
6. integration;
7. installation;
8. revision, enhancement and maintenance.

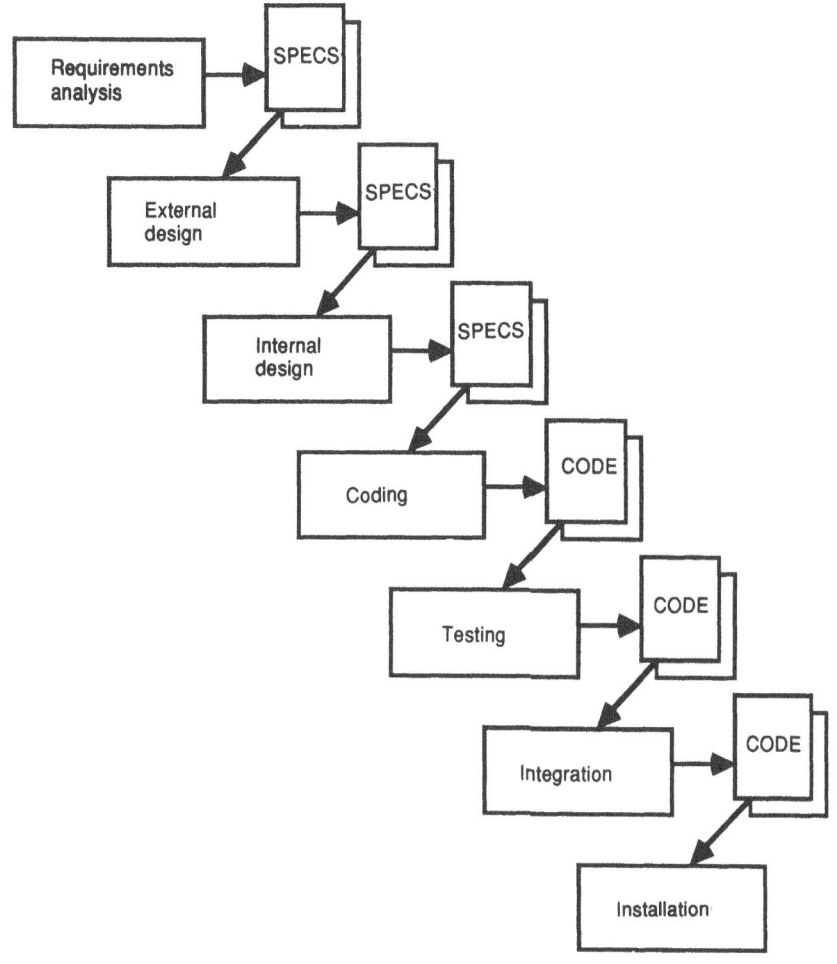

Figure 10.1 *Traditional software development lifecycle phases and deliverables.*

One important implication of such models is that management expects each phase to be completed in sequence with tangible deliverables at the end of each phase that serve as starting points of the next phase. For instance, the requirements analysis phase is expected to produce a requirements document that serves as a starting point for external design. Similarly the software system after installation serves as the basis for the revision, enhancement, and maintenance phase. In a manufacturing firm, senior managers may not be directly concerned with the technical details of the software development project. But they monitor a project's schedule and costs against such perceived models. According to traditional cost models of the software development process (Boehm, 1981), there is an extremely high cost associated with iteration. Returning to a previous phase is considered a retreat necessary for recovering from some mistake committed earlier in the project.

In contrast, we have found that the project lifecycle using the object-oriented method is more distributed and highly recursive in terms of products and activities. Progress is marked by concurrent refinement of products that would have been produced in sequence if traditional software development methods had been used. In sharp contrast to the traditional Waterfall Model of the software lifecycle (Boehm, 1981) where integration is a distinct phase towards the end of the lifecycle, object-oriented development features *continuous* integration. The fundamental paradigm-shift that needs to be realized is that the development of complex software applications require a mutual learning between the software developers and the end-users. It is not enough to study user needs up-front and then let the software developers develop in complete isolation. Prototyping encourages closer interaction with the user. User and system requirements are fully understood after iterating through several development cycles.

At each iteration, the object-oriented software development project lifecycle does break into four phases:

1. analysis
2. design
3. synthesis
4. evolution

However, the boundaries between each completed phase are much fuzzier. Within each phase, activities of other phases will be performed recursively (see Figure 10.2). Furthermore, once a project gets off the ground, it is in the constant meta-phase called 'evolution'. The most important payoff of good object-oriented implementations is that they facilitate change naturally. Within each iteration of a higher level phase, there will be another analysis, design, synthesis and evolution. There are no waterfalls, and no 'transitioning into design' from analysis (Coad, 1991). Today, authors of old software development models are rushing to revise their own models (Boehm, 1988; Coad and Yourdon; 1991). In our own experience with manufacturing applications, we have found the 'analyze a little, design a little, and prototype as necessary' strategy to be quite practical.

There is a very familiar saying about traditional software projects: 'they very quickly get to the stage of being 95% complete; however, it takes forever to complete the last 5%.' This is because traditional development technologies are very brittle against changes. There is a high cost for modifications and enhancements towards the end of a development project. We were pleased to find that object-oriented methods did indeed facilitate incremental change. The distinguishing characteristic of object-oriented development is that the time between successive iterations is minimized, thus making it possible to follow a continuum of evolution. Usually there is a high cost for the initial learning curve required to develop an architecture and a critical mass of foundation object libraries. But once a robust architecture and library have been established, the development time and cost for new applications is greatly reduced. Figure 10.3 contrasts the

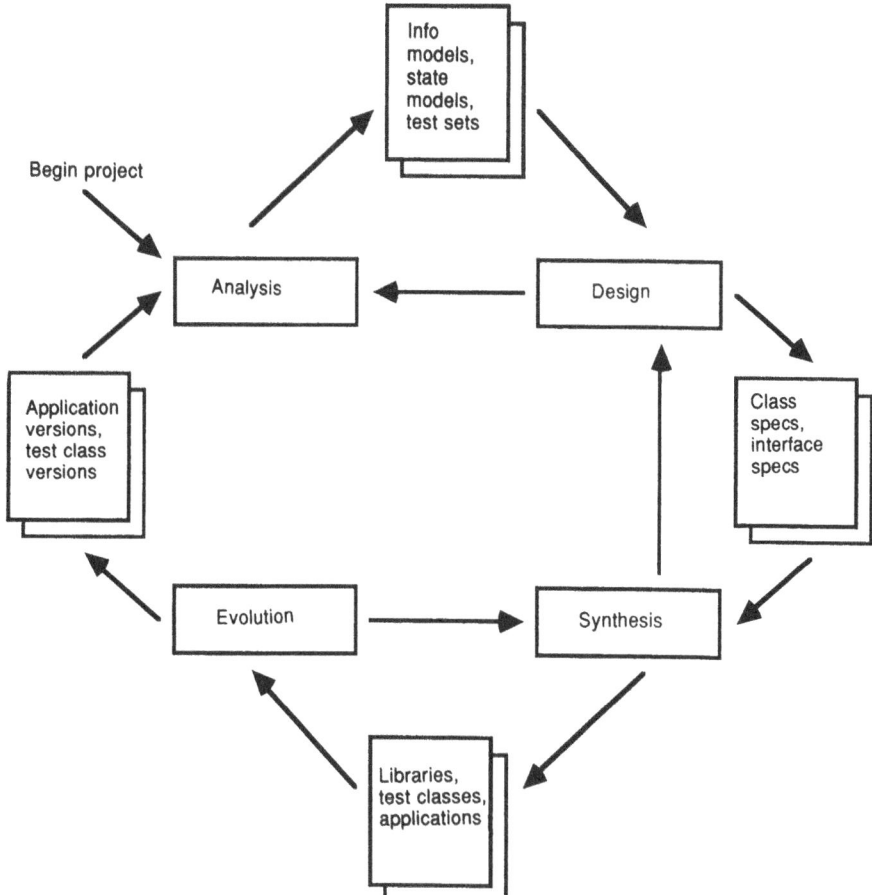

Figure 10.2 *Object-oriented software development lifecycle phases and deliverables.*

project completion profiles of the object-oriented development project with the traditional one.

10.2.2 Project tasks breakdown structure

The software work breakdown structure deals with the decomposition of the project along two views: the products and the activities.

We believe that the distinguishing characteristic about the products of object-oriented development is that each object in the library is itself viewed as a product. The evolving library and the protocols for assembly of various objects are also products. The accompanying test objects and test sets are critical products as well. Given all these, the prominence of the actual application assembled from these objects is relatively reduced. Our focus has shifted to treating the tools of making applications as the strategic product rather than the applications themselves.

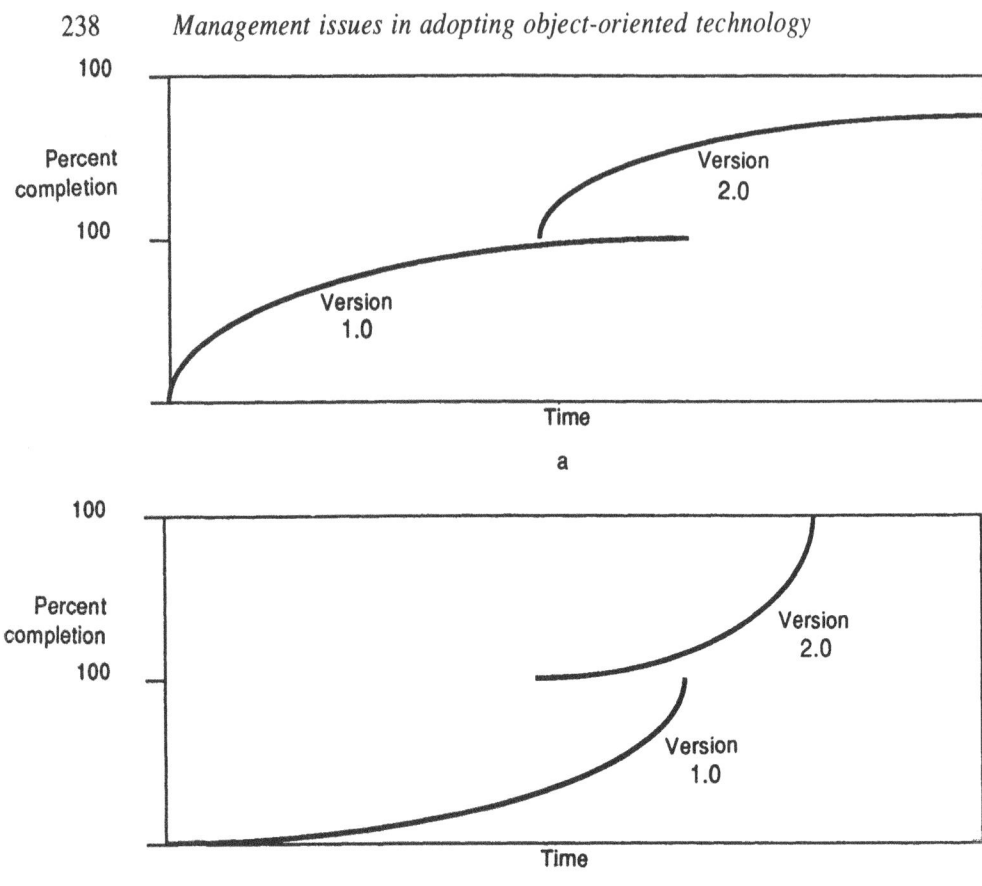

Figure 10.3 *Comparison of project completion curves; (a) traditional software development, and (b) object-oriented software development.*

Next, in identifying major types of activities, we believe that the significance of code development for the application diminishes relative to the activities of analysis and design. This is why we like to replace the term 'programming' with 'synthesis' in considering project activities. The long-term focus is on analysis and design, not synthesis. This is particularly true as analysis and design are steering towards a common notation (Coad, 1991), thus easing the way for computer-aided software engineering (CASE) and automatic code generation.

Given these considerations, the work breakdown structure in object-oriented development can be viewed roughly along the following lines:

Products:

1. class hierarchy;
2. versions of objects;
3. assembly protocols (object interfaces and system architecture);
4. test objects, and level of coverage;
5. versions of assembled applications.

Activities:

1. analysis;
2. design;
3. synthesis;
4. configuration management and quality assurance;
5. documentation;
6. project management.

According to our experience, object-oriented development is characterized by a non-traditional learning curve and more parallel distributed task-milestones. There is less effort on syntax and implementation and more effort on design. There is a need for a much greater interaction between project managers, designers, programmers and end-users. The benefits of greater interaction are obvious. Involvement breeds commitment. It also means that all parties will climb the project learning curve together and have a greater understanding of mutual limitations. In fact, with better CASE tools and the so-called 'fourth generation' languages, the distinction of the designers from the programmers has been blurred. Also, we are increasingly hearing about companies moving towards an 'empower the user' philosophy. In future, even the distinction between the designer and the end-user might diminish as well. In an ideal scenario, the role of the Information Systems (IS) department would shift from code generation and maintenance to more of a role of software development quality assurance, software engineering, tools generation and expert consultants. The IS staff will then consist largely of architects and strategists rather than programmers.

The following issues are important in changing the work breakdown structure when adopting object-oriented development technologies:

1. More attention needs to be given to system architecture. Placement of new classes in the hierarchy is far more significant a task than mechanics of installing a subroutine in a library.
2. The big picture needs to be better understood. Design decisions have to be made giving considerations to strategic needs. The current project objectives need to be understood within the overall software product strategy.
3. There is a need to budget for tasks for redesign and rework of existing classes. The designer cannot always anticipate that a component will be needed elsewhere.
4. Due to increased need for learning, there is more emphasis on prototyping tasks. However, we need to be careful not to do prototyping in lieu of design!
5. It is better to have a hardware-like view of software components. Object-oriented design results in small, loosely coupled pieces of work. Developers are more effective when they are working on clusters of related components. The emphasis is on small module coding and testing.

The software work breakdown structure also serves as a basis for estimation, collection and reporting of project cost data. Lenzi (1990) writes:

'The estimator must suppress the temptation to recalibrate for use with OOPS the traditional estimating techniques which are based on counting program modules, assigning complexity weights, and multiplying by units of average time per module needs. The historical (experiential) database of project estimates versus the actuals one may have been building is not directly transferable. A new database must be established for post-learning-curve OOPS projects.'

The traditional view of treating lines of code as a metric of progress is not applicable in developing object-oriented software. We hope that as standard work breakdown structures are defined for object-oriented development and organizations collect software costs according to these standards, they will accumulate an extremely valuable database on their software cost distributions.

It is important to note that successful adoption of object-oriented development technologies typically does not result in lower operation costs and maintenance expenditures. Organizations taking full advantage of the increased development capacity resulting from exploitation of object-oriented technologies simply end up supporting more systems. The maintenance effort per system decreases, but not the overall maintenance costs. We have found that when end-users realize the increase in development flexibility and capacity, the number of software service requests increase to fit (or exceed) the development capacity. Performance of such organizations needs to be measured in 'bang for the buck' criterion rather than reduction in overall spending.

10.2.3 Library development

Object-oriented development emphasizes the components that make up the application rather than the application itself. Object libraries are analogous to the Lego sets. Just as a child can build Lego structures from assembling different Lego blocks, a developer can build applications from libraries. However, object libraries need to be carefully planned. By just throwing together a hodge-podge of objects from various developers into a common database, one does not get a very useful library. A library should be viewed as an application assembly toolset. Just as a carpenter's toolset contains tools for different types of work at various levels of specialization, a software developer's toolset needs to support multiple levels of abstraction and various types of applications. It is typical to expect a lot of revision in a library class early on. As the classes mature and establish their place within the toolset, their interfaces and methods become rather stable. In our code documentation we found it useful to indicate the developer's subjective opinion on the level of maturity of the object. We chose to adopt the documentation convention used for BLOCS/M (Glassey and Adiga, 1989) which defines 'maturity index' values: infantile (untested, not all methods are implemented), adolescent (every method has been tested by the author), adult (used by at least one programmer besides the author), mature (extensively used by many programmers; has been subjected to extensive code review), and obsolete (replaced by new and better version).

Our software object library includes:

1. object classes specific to the manufacturing domain;
2. object classes specific to the simulation domain;
3. object classes for user interfaces;
4. parameterized utility classes to support commonly used data structures;
5. object classes for file input/output and parsing.

Once a rich library of classes is available for a domain, one has the ability for rapid prototyping for specific applications. We have found rapid prototyping with object-oriented software to be an excellent way to obtain early user-feedback and test feasibility (Adiga, 1990). However, one must be careful not to lose sight of the objective of prototyping. One should be prepared to dispose of a prototype and start from scratch. In manufacturing organizations, since the focus is not on developing software, it is sometimes hard for managers to see the utility in starting from scratch when the prototype seemed to them to be a functional starting point. For this reason prototypes need to be clearly distinguished from early versions of the actual application within the project plan of deliverables.

10.2.4 Software testing

According to Brad Cox of Stepstone Corporation, software testing is perhaps the biggest hurdle in the path of the imminent software revolution. Although software engineering professionals are still struggling to develop better methods of software testing, we found the following guidelines from Stepstone Corporation (1988) to be quite useful:

1. All modules in a software system should be tested. The basic module in object-oriented software is the 'class'. Classes can be used to test classes. Each class that is developed should have a corresponding test class.
2. The target class name should not be coded into the test class. This allows reuse of methods in test-classes written for superclasses when writing test classes for subclasses.
3. Corresponding to each method in the target class, there should be a method in the test class.
4. Each standard protocol that will be implemented in several classes should have its own test class.
5. Each test should be completely independent of all other tests. It must do its own initialization and creation of all objects needed to test the method.
6. Each test should leave the system memory unchanged.

Testing requirements and test sets need to be considered in the beginning with system requirements and interface definitions. As system requirements are refined, testing requirements are also refined – thus improving testing coverage.

10.3 Managing reusability

Reuse is made possible in two ways by object-oriented technology. Firstly, there is direct reuse of software libraries. Secondly, since objects encode behavior in addition to data, objects and object hierarchies in domain specific libraries can become a rich store of domain knowledge. This domain knowledge can be expanded and refined over time.

Reusable software is not written overnight. We have found that it typically takes several iterations before classes in a library are truly reusable. The ability to rapidly prototype using object-oriented technology can help identify reusable components faster. However, management should avoid short-term expectations for re-usability. Skills in designing for reuse must be trained. Provisions must be made for both the learning time and resources needed to foster reusability. Managers need to know what attributes distinguish reusable software from just good software. Furthermore, they need to know what metrics they can use to quantify these attributes for target setting and tracking performance.

10.3.1 Attributes of reusable software

McClory (1989) suggests the following attributes are key to reusable components:

1. understandability
2. orthogonality
3. adaptability
4. portability
5. fitness for use

Understandability

Understandability of a software component is a prerequisite for its correct reuse. In our own experience, we have found it much easier to explain and document software in the manufacturing domain when object classes have been used to represent tangible entities in the real world such as lots, machines and workers. Nevertheless, software is still only an *abstraction*. Development of understandable abstractions is made easier in the object-oriented software but it is still a skill that needs mastering with experience. Assumptions necessary for proper behavior of a component need to be made explicit in the documentation. For this reason (among others), we have found it very useful to develop documentation deliverables concurrently with software.

Orthogonality

Orthogonality is context independence of a software component. Context dependence requires the context to be transported with the component if it is to be reused. Type-checking is a common example of 'context validation'. We have found the ***assert*** command in C to be also quite useful for context validation in C++ and Objective-C applications. Context validation minimizes reuse of a part in an inappropriate context. Parameterized programming can also facilitate orthogonality. However, in most object-oriented languages, there is a penalty in performance if parameterized classes are to be used. We have found it useful to write parameterized code generators instead. A parameterized code generator would create code for specialized subclasses where the parameters (such as data types) become part of the generated code. This code can then be compiled as is, or extended for additional functionality.

Adaptability

Adaptability is the ease with which a software component can be forced into a new context. Adaptation involves changes in source code. Adaptability of a component depends on two things: readability and structure. Readability is a prerequisite to adaptability. Structure impacts the ease at which a component may be adapted. Encapsulation, when used as part of good object- oriented design helps reduce the so-called 'ripple effect.'

Portability

Computing hardware is changing rapidly. Since it is cheaper to buy new machines than to rewrite major software applications, the software must outlive the hardware. Portability is the ability to reuse a software component on a different hardware platform.

We view portability at three levels:

1. Software is *binary compatible* across machines. Programs and libraries do not need to be recompiled on a the new machine.
2. Software is *source-code compatible*. File transfer capability exists, and software can be complied on the new machine without any source code modification.
3. Software is *language compatible*. File transfer capability exists. Code in the programming language to be used on the target machine can be generated by converting source code. The conversion could range from a fully automated conversion through a fully manual one.

So far, we have not seen binary compatible code for any two different hardware platforms. Writing software that is source-code compatible is non-trivial and needs to be done as a discipline. For instance, writing C code that is source-code compatible across platforms common in the manufacturing arena is quite difficult. Even when language vendors tout source-code compatibility, we have frequently needed to require some modifications to source code when porting across machines. Frequently, functions used in code must be limited to an accepted standard subset of that available in the language. Staying up-to-date and active in standards activity helps in the awareness of dominant trends. In addition, through the use of encapsulation provided by the object-oriented paradigm, machine dependent code can be localized to implementation specific layers.

Fitness for Use

In addition to all four attributes discussed above, cost and reliability are important criteria in considering fitness for use. It does not make sense to reuse a component when it is more expensive to find the software, understand it, adapt it, and port it, than to reengineer it. This is why management edicts ordering reuse of mandated libraries does not always result in gains in productivity. Actual coding typically takes less than one-fifth of the total development time. The real savings in reuse are in reusing not just the code, but the design and development effort that went into it.

Another important feature that makes software fit for reuse is 'scaffolding.' Scaffolding is the built-in-self-test within software modules. Software testing has been discussed previously in this chapter. It is important to note that the test objects and test-sets are part of the delivered software library. Scaffolding can be used to confirm that the component is working as intended.

Object-oriented technology provides powerful tools for reusability through inheritance and polymorphism. However, proper use of these features comes only with experience. Reusable software must balance generality with specificity. It is equally impractical to carry unnecessary baggage in a single general component as it is to have too many of specific components. We have had considerable success with following the philosophy: 'design objects so that simple things should be easy to do, and difficult things should be possible.'

10.3.2 Metrics of reusability

The problem with traditional software reusability metrics is that there are no hard measurable metrics that are easy to track. Available qualitative metrics are good for setting targets; but it is impossible to determine how well an organization is performing against these targets. McClory (1989) reported on four different types of reusability metrics:

1. code-oriented metrics
2. system structure metrics
3. readability metrics
4. portability metrics

Code-oriented metrics include lines of code, volume, and cyclomatic complexity. System structure metrics deal with information flow, package unity and component access. Readability metrics are assigned using a subjective scoring of code features. Portability metrics rate the 'transportability' and 'migratability' of software systems. Transportability depends upon the availability of compatible peripherals on both systems so that files can be transported across. Migratability is the effort needed to make the transported files useful.

What seems to be the basic problem with such metrics is that they attempt to measure reusability by measuring from the code itself, instead of measuring the extent of its reuse. From a management perspective, one would like to have metrics such that targets can be set and actual performance monitored against them. Thus, when considering objects in a library, measurement of actual reuse is considerably easier. For instance, one can simply count the number of applications that use a given object in a library. In addition the use of individual methods within an object can also be counted. However, we have found it of little benefit to measure reuse at too detailed a level. We tend to measure reusability of the library as a whole. It is of interest to us to know what level of effort will be needed to develop new applications (or enhance existing ones) given the library. We view reuse at four levels:

1. direct reuse of objects (without subclassing);
2. reuse of objects by subclassing;
3. reuse of part of the library while replacing some classes;
4. indirect reuse through reuse of models.

We do not think that it is realistic to expect a library to be so general that every application in the domain can be assembled by direct reuse. The 'simple things should be easy to do, and difficult things possible' philosophy helps decide on the appropriate level generality. The ability to subclass is one of the key advantages of object-oriented programming. If application developers frequently need to replace some classes in order to build their applications, it is likely that the library (or just those objects) are poorly designed. We have also found indirect reuse of tremendous advantage from time to time. Indirect reuse can be achieved through reuse of data models or behavior models or both. For example, in one case we used a common factory technical database to provide input to a number of CIM applications. In addition, the behavior and decision-making model encoded in the objects for a simulation-based scheduler was reused in a capacity analysis tool that used a conventional simulation language (Najmi and Stein, 1989).

Thus, the attributes of reusable software, discussed earlier in this section, apply just the same to object-oriented programming as conventional programming. We have found that software that is written with both a good understanding of the domain and good grasp of object-oriented design is more likely to be reusable.

We feel it important to keep a long-term view in promoting reusability. Reuse will happen naturally if software is written that is worthy and easy for reuse. Forced reuse by edicts will result in a progressive degradation of overall software quality. For this reason, we suggest that when building new applications (or objects) one should always start with the real problem to define the requirements on the software. It is best to define these requirements in the terms of the problem domain *without making assumptions about the library.* After the requirements are clearly and objectively defined, we look in the library to see if there are objects that come close to meeting the data and behavior requirements and could be reused for an actual implementation. If such objects exist, inheritance can be used to specialize these objects as needed. However, if no objects address the needs adequately, it is better to write new software. It is very costly in the long run to keep convoluting the problem specification with patch-ups just to make it fit existing but unsatisfactory solutions.

10.4 Resourcing and staffing

10.4.1 How much?

We agree with Booch (1991) that, compared to traditional lifecycle models, object-oriented technology requires about the same human resources for analysis, but significantly more resources for the design phase. Coding and testing phases require relatively less resources. Even fewer resources still are required for integration. Booch claims that the net sum of all human resources required for object-oriented development is usually equal to or less that required for traditional approaches, and the resulting product tends to be of far better quality. We would add a reminder that there are qualitative differences in the type and skill of people needed for object-oriented projects. This is discussed next.

10.4.2 Who?

Staffing up for object-oriented technology requires people with skills in the area of system-architecture, class design, application design, and object-oriented programming. Good people are hard to find. There are very few people with both the up-to-date knowledge of the technology and experience developing industry strength systems. Lenzi (1990) suggests that it is more critical to find people with flexible minds rather than those with extensive language background. People who are able to change and adapt, those who can view mistakes as vital to the learning process, are best suited for making the change from procedural to object-oriented development. In our own experience, we have had success with people with good design skills that are familiar with programming. Programming languages can be learned, but there is no substitute for poor design – in any technology.

10.4.3 Organization

Another issue that must be decided early is the placement of the development group within the organization. New technology introduction requires an appropriate managerial and technical umbrella to see it through the initial stages. As we will discuss in section 10.6, management must allow for a significant learning curve and a large number of unknowns at the onset of the project. It is important that developers have access to technical people that are experienced in the technology. Formal technical support can be set up through employment of consultants. Informal support can be through electronic news groups, user groups, interactions with university contacts, and cooperation with other groups within the company. Technical conferences also provide an early information exchange forum for emerging technologies.

Management must be more careful when undertaking the first software development project in manufacturing using object-oriented technology. They should avoid the temptation to treat it as just another software development project. In most companies software for manufacturing is supported by IS organizations. However, the culture of most IS organizations is built around mature, tried, and tested technologies. There is little room for exploration and research. It discourages risks involved with new technology introduction. Research and Development (R&D) organizations are better suited in such cases.

Our own first project was initiated from within an Advanced Development group. Advanced Development groups routinely engage in treading new ground by taking up high risk and high reward projects. Once we gained experience with the technology and confidence in its products, an Advanced Manufacturing Engineering group was established within the manufacturing organization with a charter to exploit emerging technologies for manufacturing. While this group remained in close collaboration with our IS organization throughout, it served to fill this distinct role of providing technical and managerial support to the introduction of new technology in manufacturing.

10.4.4 Expectations

Commitment to projects should be made only after a clear understanding of development priorities and capacity. McCoy (1989) writes:

> 'We are all familiar with what happens when the memory demand on a shared system greatly exceeds its capacity. The processor remains busy, but throughout drops. This is *thrashing*, and it has a direct analogy to an overcommitted engineering organization with more projects than capacity.'

It is important not to raise management's expectations to unreasonable levels in an effort to sell object-oriented technology. Furthermore, restraint should be used in committing to too many development projects once the productivity advantages of object-oriented technologies are achieved.

10.5 Supplier management

In our view, a supplier within the object-oriented development context is any party that supplies products that influence the overall software product. This includes software language vendors, foundation library vendors, suppliers of domain-specific object models, architectures, and other deliverables. In fact, different teams of developers can be viewed as suppliers to each other as their products all need to come together in an integrated system.

The goal of supplier management should be 'drop-in' components. The interfaces and test requirements between components need to be so clearly defined that once the components have been tested against them, they drop into the overall system with no surprises. Black-box views of the rest of the world as viewed from each object class can be constructed. Simulations can be developed to emulate the expected behavior of external objects. These essentially serve as stubs until the actual objects are dropped in as their versions are released from their suppliers. Similar to the ship-to-stock philosophy for vendor supplied material in manufacturing, the need for modification of supplied software before use should be eliminated. Ship-to-stock in software implies a partnership between the supplier and the customer so that bugs in the supplied software are fixed before delivery. Similarly, suppliers get greater visibility of changing needs of the customer so that they can accommodate these changes in a timely manner.

Changes in the software architecture can likely change the requirements between suppliers and customers of software components within the architecture. All software architectural decisions need to be simulated. Frequently, architectural changes have far-reaching but poorly understood impact. Data gathered from a small investment in simulation and development of throw-away rapid prototypes can help avoid huge costs down the road.

10.6 Planning and budgeting

10.6.1 Planning for the long-term

We believe that the true payoff from investing in object-oriented technologies is realized in the long-term. Adoption of object-oriented technologies therefore needs to be viewed as a strategic investment. Emphasis needs to be placed on planning for the long-term based on the needs and expectations of the organization.

One implication of this is that the development organization needs to have a well-defined mission consistent with the mission of the company. The mission of the software development organization serves as a bottom-line criterion in making difficult tradeoffs in strategic decisions.

Planning for the long-term means that returns on investment from alternatives ought to be based on comparisons over longer horizons. Unfortunately, the technology is new and there is not enough data accumulated to make crisp cost estimates too far out in the future. Organizations frequently end up still needing to justify capital appropriations requests for such projects in traditional terms. We have found a compromise approach where justifications are based on both strategic and return-on-investment criteria.

When object-oriented technologies are being introduced in an organization, special attention needs to be given to issues of training, methodology changes, and learning curve in the project plan. Budgets must be allocated for training, software development tools, developer resources, and external consulting support. The transition into object-oriented technologies involves higher initial costs.

10.6.2 Decision schedules

Traditional project schedules itemize tasks and milestones in a definite progression. The object-oriented development lifecycle is naturally iterative. Decisions made at each iteration can impact the overall course of the project. Also, because decisions are to be more numerous and frequent, timely decision-making is critical. Without special attention to decision points, object-oriented software development projects stand the risk of being bogged down in decision paralysis. We have found decision schedules to be extremely useful in this regard. The key idea behind the concept of decision schedules is to anticipate opportunities for decisions well in advance and develop plans that are robust against the possible set of decisions. Our plans must answer the following questions:

1. When will the decision opportunities arise?
2. When will the decision opportunities lapse?
3. What information will be critical to these decisions?
4. Will this information be available within the decision opportunity window? What plans and contingencies are in place to ensure timely availability of this information?
5. What is the range of possible decisions? What decisions are most likely?
6. How robust is the project plan against these possible decisions?

Thus, the decision schedule should resolve almost everything but the decision itself, ahead of time. When the opportunity arrives and the relevant information is available, the decision is made briskly and the project moves on. Use of decision schedules makes it possible to avoid getting the project hung up in decision holes at each iteration.

10.6.3 Integration-driven scheduling

While developing object-oriented systems that need to be integrated with existing facilities such as databases, other applications etc., the schedule should be driven by integration needs. In planning an integration-driven schedule, one needs to know the key components to be integrated, and the time it takes to develop them. In contrast to typical project schedules that work forward from milestone dates, integration-driven schedules are constructed by identifying integration points and working backwards from them (see Figure 10.4).

McCoy (1989) discusses integration-driven scheduling as a key tool to achieving 'just-in-time engineering' at NCR. He points out that integration-driven scheduling allows for the maximum possible development and testing that can be performed before the components are delivered. The time to develop system components can be minimized by exploiting the maximum amount of parallelism possible in the application of development resources. He notes that integration-driven scheduling would degrade to end-date planning if one fails to anticipate integration dates by more than the development time. Integration-driven scheduling requires marshaling of resources prior to deliveries, improved supplier management, careful external schedule tracking, and increased advance planning. Integration-driven scheduling ensures that all items are available at the key integration points.

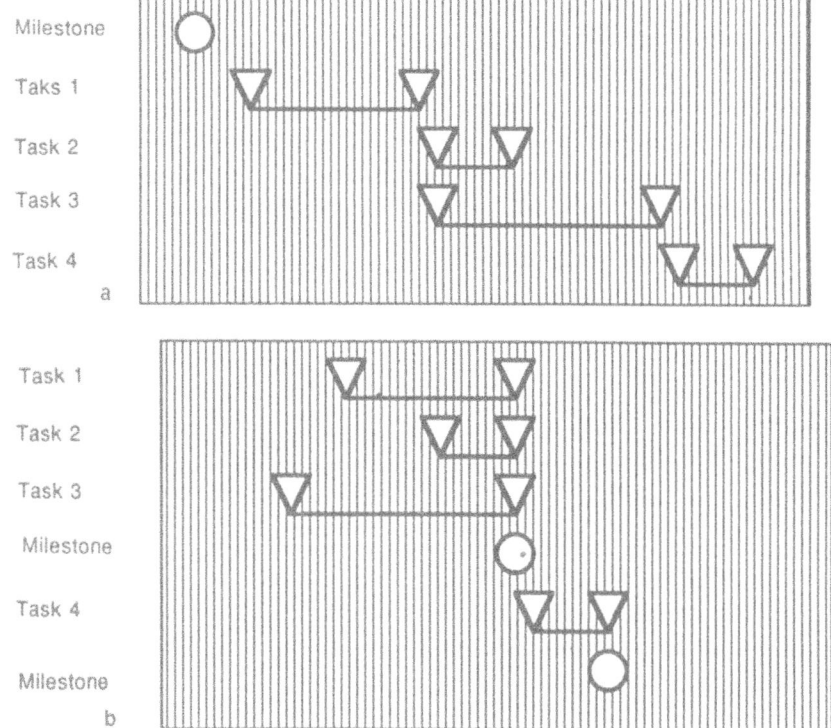

Figure 10.4 *(a) Traditional project scheduling; (b) Integration-driven scheduling.*

10.6.4 Orchestrating an integrated management plan

A typical detailed project plan lists project milestones and the tasks necessary to achieve them. The completion schedule and ownership responsibilities of each task are identified. Resources are also allocated for each task. In addition, we have found it critical to the success of large scale object-oriented development projects to have an 'integrated management plan' that deals with the high degree of interdependency between concurrent tasks which is characteristic of this paradigm. The integrated management plan is monitored by a inter-departmental task force that acts as a communication link between the organizations involved. Individuals in the task force need to be in decision-making positions within their respective organizations. Their close communication leads to continuous refinement of sub-program plans as the interdependent project components evolve. These task forces need to be established at the onset of the project, well before development of individual components is started. The task force needs to stick with the project from inception to completion. Orchestrating such projects is hard work. The planning and project management responsibility cannot be treated as a part-time job.

10.7 Conclusion

We believe that adoption of object-oriented technology requires a long-term view. It is a strategic investment that yields returns over the long haul. Object-oriented software development projects have a different lifecycle. The way software development projects are managed must be changed to allow for a high level of concurrency and just-in-time engineering. The focus of management attention shifts from the software that makes up the application, to the software libraries that makes rapid application development possible. The project work breakdown structure changes and traditional estimation formulas do not apply. Integration is continuous throughout the project rather than a separate phase towards the end. However, once object-oriented technology is successfully adopted, we find it is easier to keep pace with changing user needs.

References

Adiga, S. (1989) Software modeling of manufacturing systems – a case for an object-oriented approach. *Annals of Operations Research*, **17**, 363–378.

Adiga, S. and Lin, W. T. (1990) An object-oriented architecture for production scheduling systems. Technical Report No. ESRC 90–1, Engineering Systems Research Center, University of California at Berkeley.

Boehm, B. W. (1981) *Software engineering economics*. Prentice Hall, 34 164.

Boehm, B. W. (1988) A spiral model of software development and enhancement. *IEEE Computer*, May, 61–72.

Booch, G. (1991) *Object-oriented design with applications*. The Benjamin/Cummings Publishing Company, Inc., 197–219.

Coad, P. and Yourdon, E. (1990) *Object oriented analysis*. Prentice-Hall, Englewood Cliffs, NJ, USA.

Coad P. (1991) OOA and OOD a continuum of representation. *Journal of Object-Oriented Programming*, February, 55–56.

Glassey, C. R. and Adiga, S. (1989) Conceptual design of a software object library for simulation of semiconductor manufacturing systems. *Journal of Object-oriented Programming*, Sept./Oct.

Lenzi, M. A. (1990) Adopting object-oriented techniques: management and organizational impacts. *Hotline on Object-oriented Technology*, January, 1–6.

McClory, T. J. (1989) Measurable attributes of reusable software: a survey. *The NCR Journal*, December, 14–23.

McCoy, C. (1989) Just-in-time engineering. *The NCR Journal*, December, 2–13.

Najmi, A. and Lozinski, C. (1989) Managing factory productivity using object-oriented simulation for setting shift production targets in vlsi manufacturing. *Proceedings of the Autofact Conference*, Society of Manufacturing Engineers, November, 3–1: 3–14, Detroit, MI.

Najmi A. and Stein, S. (1989) Comparison of conventional and object-oriented approaches for simulation of manufacturing systems. *Proceedings of the IIE Society of Integrated Manufacturing Conference*, Atlanta, GA, November.

Stepstone Corporation (1988) *Programming in Objective-C*, Course Manual.

11 Concluding remarks

S. ADIGA

The methods and toolkit used to build software components of current CIM systems seems to be woefully inadequate. Object-oriented computing offers new directions to open up a new class of flexible and robust solutions. This book contains a discussion of basic concepts, the techniques needed to design OO systems, a proposal for a new architecture, and examples of some significant practical systems. We have addressed the technical as well as organizational issues of building OO systems for manufacturing applications.

We have presented a model of an architecture for CIM systems based on the following premise: information is one of the few links among various activities in manufacturing, and a common 'information model' can be used to act as the reference for designing and implementing CIM applications. We have also discussed how this architecture may be built as collections of loosely coupled, interacting self-contained software objects. Our approach bridges the gap in representation between the user interface object, the simulation object, and the manufacturing entities in a plant. This framework has the potential to eventually lead to an open systems architecture for the CIM systems of the 21st century.

While we have taken an enthusiastic attitude towards the OO approach, we would like to remind the readers that it is no panacea! Good software design needs more than tools and techniques. It needs knowledge, skill and discipline (Constantine, 1990). Buying into object-orientation is not a substitute for the fundamental skills required of a software engineer.

11.1 Where is the OO industry headed?

Object-oriented technology is coming out of the closets of programmers who covet esoteric techniques, into factories where it improves the productivity of manufacturing systems. Market studies show that object-oriented software products are poised for growth in the 1990s. According to a study by Ovum Ltd., a market research group based in London, total annual revenues for OO development tools and languages is expected to reach nearly $2 billion by 1996. During the same period, the OO products marketed to end-users is expected to yield around $250 million.

Perhaps the database is the most developed area in terms of choice and power of products available. We already see all the makings of a war between those supporting relational databases and those who favour the object database. It is virtually a repetition of the past when the relational model was pushing out the hierarchical/network models. However, relational DBMS have not pushed older models out of existence; similarly, all the data models will co-exist up to a point which will be determined by the market forces. The most likely scenario is that the RDBMS and OODBMS will come to co-exist in a distributed network environment, each doing what it does best. In a distributed database environment, transactions could occur where they are most natural and appropriate.

In spite of these optimistic projections, there are still obstacles to be overcome. A lack of standards could affect compatibility among the class libraries. An encouraging trend is set by the Object Management Group (OMG), of Framingham, Massachusetts, USA. OMG is a not-profit association of vendors and users of OO tools and systems (totaling around 120 in June 1991). It is an international group, although the majority of members are from the USA.

11.2 Growth in the manufacturing area

We share Brad Cox's conviction that the software revolution is unlikely to be driven by the software development community, just as the industrial revolution was not driven by the cottage industry gunsmiths. While the computer scientists are working on extending the scientific underpinnings of using object-orientation, and creating powerful tools to build OO applications, engineers need to work on the next generation of applications.

Growth in many areas is required to support the concept of building applications by assembling objects bought from vendors. Initially, we can expect more firms like Consilium and Savoir to build large customer applications from their library of objects. As the market matures, we are likely to see companies selling object libraries of generic manufacturing objects for different sectors of manufacturing such as electronics, aerospace, the automobile industry, etc. In such an event, the price of software could come tumbling down.

We are also aware that many manufacturing companies have in-house projects to build OO applications. This trend is also fuelled by the growth in availability of consulting firms who specialize in OO methods, such as Rational Consulting or Project Technology of Berkeley, and systems specialists such as Coad and Yourdon, Colbert, and others.

The awareness in the manufacturing community (academic and industrial) is seen in the encouraging response to the First International Conference on Object-Oriented Manufacturing Systems (ICOOMS), to be held in Calgary, Canada. The conference is being organized by a group headed by Professor D. Norrie of the University of Calgary, Canada.

Dealing with common objects viewed with the same philosophy may lead to the development of a commonly conceived manufacturing interface that relates to the way people perceive machines, parts, etc.; that is, based on a common set of objects. This has the potential to have every manufacturer's software run the same way, thereby cutting down learning time for the users. This is not an unrealistic picture, as we have seen similar things happen in the past. We do not have to take fresh driving lessons when switching automobiles from, say, a Toyota Corolla to a Ford Escort or vice-versa.

11.2.1 Manufacturing as a distributed computing system

One of the most promising areas for future developments in the application of OO concepts in manufacturing is in viewing it fully as a distributed computing system. While we have used some concepts from distributed systems research in our proposed architecture, there are many more interesting issues to be explored.

Distributed computing applications can be grouped into two categories. Those in the first category include a process that requests service of a remote process (or processes), as this service is not available locally. These applications are well suited by a client – server model where a client's request causes some processing, and a result is returned with a message from the server to the client (Christiansen and Tanik, 1988). This model is useful for retrieving information that is stored in a remote location or sharing of peripherals.

The second category consists of applications where a set of processes cooperate in the solution of some problem. One hypothetical application belonging to this category is described next, emphasizing its potential for improvement of an existing problem.

Batch processing on a mainframe computer is a common practice for manufacturing firms that run MRPII software. Usually, the MIS departments run these MRPII programs during weekends or late at night to minimize computing load. This practice introduces a lag between the data used by MRPII and other shop-floor management systems which, in turn, leads the shop-floor personnel to come up with their own plans or schedules. Additionally, it also may introduce inconsistency between the two systems.

In a distributed system environment, several personal computers or workstations collect, monitor and control some aspect of factory production. They communicate with each other as required. They run concurrently and independently except that they share common data.

In a workstation-based scenario of computing, local plants will do their own planning and forecasting and these forecasts are sent to a mainframe system to be aggregated. Such an aggregated plan will allow the firm to take advantage of volume discounts for raw materials and supplies, etc.

Solberg (1989) identifies some research issues leading to a decentralization of planning and controlling of production. They include new themes in cooperative systems, heterarchy, information encapsulation, real-time negotiation of resources and a mechanism for automatic learning.

Designing manufacturing systems as distributed systems gives rise to some interesting problems. Jablonski *et al.* (1990) discuss an approach for dividing all manufacturing applications into global and sub-applications. An OO view of the world helps promote distributed thinking through flexible encapsulation. Its different approach to synchronization and concurrency may be exploited to an advantage.

The trend in most distributed systems research appears to be in the direction of building transparent, seamless, integrated environments. Most of the strengths of OO come from its ability to model the real world accurately. As we know, the real world is full of conflicts (manufacturing is no exception!) and our quest for global consistency may neither be realistic nor required. Perhaps a better approach to handling these is to acknowledge the conflicts and model them explicitly rather than ignoring them or making simplifying assumptions.

It is well accepted in computer science that getting autonomous objects, programs, people, or whatever to work together is a difficult problem (Tsichritziz and Nierstrasz, 1989). A mechanism for cooperation among objects (hardware and software) and people towards a common overall goal is an important research topic in this area.

11.3 Final words

In his classic book on scientific philosophy, *The Structure of Scientific Revolutions* (1970), Thomas Kuhn described scientific theory as a paradigm controlling how we view the world. Kuhn claimed that a successful revolution in science generally takes a full generation. This is because most of the current practitioners will be reluctant to adopt a new approach.

Visionaries such as Alan Kay, K. Nygaard, O. Dahl, Adele Goldberg, Brad Cox, Bjarne Stroustrup and others have laid the foundation for a paradigm that has the promise to change the way we build software systems. But these concepts have to be nurtured, interpreted and developed further in the context of each of our application areas. We expect that this book will heighten interest among professionals in this promising technology. We also hope that our book will act as a catalyst for people who intend to apply OO concepts to build further applications in manufacturing, and thus speed up the software systems revolution!

References

Christiansen, M. G. and Tanik, M. M. (1988) Design issues for a distributed software engineering support system. *International Journal of Computer Applications in Technology.* 1(1/2), 85–95.

Constantine, L., quoted in Keuffel, W. (1990) House of structure. *Unix Review,* 9(2), 28–36.

Jablonski, S., Ruf, T. and Wedekind, H. (1990) Concepts and methods for the optimization of distributed data processing. *Proceedings of the Second International Symmpo-
sium on Databases in Parallel & Distributed Systems*, 171–180.

Kuhn T., (1970) *The Structure of Scientific Revolutions*, 2nd edn., University of Chicago
Press.

Solberg, J. J. (1989) Production planning and scheduling in CIM. Invited paper, in
Information Processing '89: Proceedings of the IFIP World Computer Congress, (ed.
G. X. Ritter), Elsevier Science Publishers, 919–925.

Tsichritzis, D. C. and Nierstrasz, O. M. (1989) Directions in object-oriented research. In
Object-oriented concepts, databases, and applications (eds. W. Kim and F. H. Lochoov-
sky), ACM Press, 523–536.

Appendix A: OO resources

Compiled by PAUL WORHACH

Industrial Engineering & Operations Research Department, U. C. Berkeley, CA

Sources for further research

The following is a summary of the major journals, conferences, societies, and other primary sources of information dealing with object-oriented programming and manufacturing systems. As with the field of OO systems itself, the sources of information are dynamic; new conferences and journals appear each year. This is not an exhaustive list; but it is intended to point the reader in the proper direction.

Journals/magazines

Journal of Object-Oriented Programming
Dr. Richard Wiener, Editor
588 Broadway, Suite 605
New York, NY 10012

Contains technical papers and regular columns on different languages, special topics, product and industry news. Published by SIGS, which also produces newsletters, reports and directories of OO systems, and sponsors the SCOOP conferences.

Object Magazine
Marie A. Lenzi, Editor
SIGS Publications

Hotline on Object-Oriented Technology
SIGS Publications
588 Broadway
New York, NY 10012

The C++ Report
Robert Murray, Editor
SIGS Publications

OOPS! The Newsletter of the Object-Oriented Programming Society
Bruce Andersen, Editor
Electronic Systems Engineering, University of Essex
Colchester, C04 3SQ, United Kingdom

OOPS Messenger
Published by the SIGPLAN committee of the ACM
11 West 42nd Street
New York, NY 10036

First Class
Published by the Object Management Group

While the above publications are written for an audience interested in OO computing, special issues have been published by the following journals/magazines on OO topics:

IEEE Software
IEEE Expert
Communications of the ACM
11 West 42nd street
New York, NY 10036
BYTE Magazine

Conferences

European Conference on Object-Oriented Programming

An annual conference devoted largely to programming and language development, but includes some application-oriented work as well.

Contact: Centre Universitaire d'Informatique
12 rue du Lac, CH-1207 Geneva, Switzerland

Object-Oriented Programming Systems, Languages and Applications (OOPSLA)

An annual conference sponsored by the ACM, this is the primary US conference for OO languages and systems.

Contact: Association of Computing Machinery (ACM)

International Conference on Object-Oriented Manufacturing Systems (ICOOMS)

A new conference devoted to manufacturing applications of object-oriented design and analysis.

Contact: Dr. Douglas Norrie
Division of Manufacturing Engineering
University of Calgary

Technology of Object-Oriented Languages and Systems (TOOLS)

This conference focuses almost exclusively on industry-oriented applications, including information and manufacturing systems. A wide series of tutorials are included to complement the presentation of original work and successful applications.

Contact: Betrand Meyer
Interactive Software Engineering, Inc.
(805) 685-6869

Professional seminars/Product exhibitions

Seminars and Conference in Object-Oriented Programming (SCOOP)
588 Broadway, Suite 604
New York, NY 10012

East and West coast and European conferences sponsored by SIGS Publications (the publisher of *Journal of Object-Oriented Programming*; they are focused on vendor products.

Object World
C/O World Expo Corporation
P.O. Box 9107
Framingham, MA 01701–9107

Trade Associations

The Object Management Group
492 Old Connecticut Path
Framingham, MA 0170
USA

An international organization of information system vendors, software developers and users founded to promote the theory and application of OO technology in software development. Publishes *First Class*, a group newsletter. This group is also working on standardization issues.

Glossary

We provide here a brief definition of some technical terms used in the book. A more complete description can be found in the body of the book.

Abstraction The process of formulating generalized concepts by extracting common qualities from specific examples (Blair *et al.*, 1991, p. 3).

Active object or Actor An object that has an independent thread of control. It can monitor events and take action autonomously.

Attribute Property or characteristic of an entity.

Asynchronous message Object sending a message that does not wait for results (i.e. return) from the object receiving the message.

Binding Logically associating a name (of a message or a class) with a location of the program code to be accessed by an invocation of the name in a program. Binding is applied in connection with messages and objects. Binding denotes the association of a name of a message with a method to execute the message, or the name of a class with objects being created. See also **early binding** and **dynamic binding.**

Class A description of objects sharing common properties and behavior. It may be viewed as a template for describing an object's protocol (methods) and structure (instance variables). A class can be used to create objects.

Client object Object issuing a request for service. See also **server object**.

Computer-integrated manufacturing (CIM) An effort to integrate the activities in manufacturing through the medium of computers. It is a strategy based on computer-based sharing of resources and the making of timely and appropriate information to optimize the performance of the entire manufacturing organization.

Concurrency The ability of a computer program to do two or more things simultaneously.

Database A collection of interrelated data.

Database management systems (DBMS) A set of software to manage the organization, storage, retrieval and operation of a database.

Dynamic binding Binding that is performed after the request for a service is issued. For example, a method is located dynamically when it is needed and binding between the message and the corresponding method takes place just when it is needed, i.e. when the program is running. See also **binding**.

Early binding Binding that is performed before a program is run. That is, the connection among the components of a program is made early, prior to the actual issuing of the request for services (of a method or creation of an object, for example).

Encapsulation Bundling of data and methods together (in an object). It also implies that access to data is allowed only through the object's methods.

Entity Something that can be distinctly identified (Chen, 1976). An entity may be broadly interpreted as a tangible object (a thing such as an automobile, or a person) or an intangible concept or event (e.g. failure of a machine, preparation of a purchase order) about which the organization chooses to collect and store data.

Entity set A collection of entities that have similar characteristics (e.g. students, operators, wafers).

Events Occurrences that cause the system to change its state. See also **system state**.

Information hiding The idea that the internal implementation-related information of an object should be hidden from the users.

Inheritance Sharing of methods and instance variables (data types) among classes and subclasses. Inheritance enables programmers to program only what is different from previously defined classes (Winblad *et al*. 1990, p. 267).

Manufacturing Message Specification (MMS) An international standard designed to support communications to and from programmable devices in a automated manufacturing environment.

Message-passing A mechanism for sending messages from an object to other objects or itself.

Method Code that may be executed when a message is received by an object.

Multiple inheritance The ability of a class to inherit data and methods from more than one class. See also **inheritance**.

Object A fundamental unit of computation in the OO world. An object is an abstraction of a real world or conceptual entity. It is implemented as a computational entity that encapsulates state (variables) and operations (methods). It responds to messages (requests for services) by invoking its methods. Objects are organized in classes.

Object-orientation Involves looking at the world as a set of objects that are related and communicate with one another by passing messages. See **object**.

Object-oriented world view   See **object-orientation**.

Persistent object   An object that outlives the process or thread that created it. A persistent object exists until it is explicitly deleted.

Polymorphism   Refers to the ability of objects to respond differently to the same message in ways unique to their respective behavior.

Protocol   A set of conventions used to exchange messages.

Relationship   An abstraction of the association between entities.

Server object   An object providing a response to a request for a service. See also **client object**. A given object may be a client for some requests and a server for others.

Specialization   A class x is a specialization of a class y if x is defined to inherit from y.

State (of an object)   One of the many possible conditions under which an object may exist (after Booch, 1991, p. 518). At any given point in time, the state of an object is characterized by the values of its instance variables.

Static Binding   See **early binding**.

Subclass   A class that inherits from one or more superclasses. The latter is true in the case of multiple inheritance.

Superclass   The class from which some other class inherits.

Synchronous message   A messaging activity where the object sending the message pauses to wait for acknowledgement of the message execution by the receiving object. Most references to messaging in this book are synchronous except in Chapter 9 (OOPS in real-time control applications).

System   A group of interacting objects. The purpose of a system is to provide the necessary response whenever an input stimulus, a time event, or an anomalous condition is recognized by the system (Kowal, 1988).

System state   One of the many possible conditions under which a system may exist.

Transient object   An object that lives only during the lifetime of the process or thread that created it.

Abbreviations (or acronyms)

Some abbreviations used and their descriptions are listed below. Their relationship to object-orientation is described in the main body of the book.

CAD   Computer-aided design
CASE   Computer-aided software (or systems) engineering

CIM   Computer integrated manufacturing
CAM   Computer-aided manufacturing
MAP   Manufacturing Automation Protocol
MMS   Manufacturing Message Specification
MRP   Materials requirement planning
OMG   Object Management Group
OOA   Object-oriented analysis
OOD   Object-oriented design
OOP   Object-oriented programming
OOS   Object-oriented software
OODBMS   Object-oriented data base management systems
RAM   Random access memory
RDBMS   Relational data base management systems
RPC   Remote procedure call
SQL   Structured Query Language (a language used to create, access, and manage data in relational databases)

References

Blair, G. *et al.* (eds.) (1991) *Object-Oriented Languages, Systems and Applications.* Halstead Press.

Booch, G. (1991) *Object-Oriented Design with Applications.* Benjamin/Cummings Publishing Co., CA.

Chen, P. P-S, (1976) The entity – relationship model – toward a unified view of data. *ACM Transaction on Database Systems* **6**(1).

Kowal, J. A. (1988) *Analyzing Systems.* Prentice-Hall.

Soley, R. M. (ed.) (1990) Object Management Architecture Guide. *OMG TC Document 90.9.1.*

Winblad A. *et al.* (1990) *Object-Oriented Software.* Addison-Wesley Publishing Co. Reading, MA.

Index

The manufacturer's authorised representative in the EU is Springer
Nature Customer Service Centre GmbH, Europaplatz 3, 69115 Heidelberg,
Germany. If you have any concerns regarding our products, please
contact ProductSafety@springernature.com

Printed and bound by CPI Group (UK) Ltd, Croydon, CR0 4YY

23/04/2026

02095623-0006